Random Fields and Stochastic Partial Differential Equations

Mathematics and Its Applications

Volume 438

Random Fields and Stochastic Partial Differential Equations

by

Yu. A. Rozanov

Steklov Mathematical Institute,
Moscow, Russia
and
CNR-IAMI,
Milan, Italy

KLUWER ACADEMIC PUBLISHERS
DORDRECHT / BOSTON / LONDON

A C.I.P. Catalogue record for this book is available from the Library of Congress.

ISBN 978-90-481-5009-0

Published by Kluwer Academic Publishers,
P.O. Box 17, 3300 AA Dordrecht, The Netherlands.

Sold and distributed in North, Central and South America
by Kluwer Academic Publishers,
101 Philip Drive, Norwell, MA 02061, U.S.A.

In all other countries, sold and distributed
by Kluwer Academic Publishers,
P.O. Box 322, 3300 AH Dordrecht, The Netherlands.

This is a completely revised and updated translation of the
original Russian work of the same title,
Nauka, Moscow ©1996 (in Russian).

Printed on acid-free paper

CONTENTS

Foreword

This book considers some models described by means of partial differential equations and boundary conditions with chaotic stochastic disturbance. In a framework of stochastic Partial Differential Equations an approach is suggested to generalize solutions of stochastic Boundary Problems. The main topic concerns probabilistic aspects with applications to well-known Random Fields models which are representative for the corresponding stochastic Sobolev spaces. (The term "stochastic" in general indicates involvement of appropriate random elements.)

It assumes certain knowledge in general Analysis and Probability (Hilbert space methods, Schwartz distributions, Fourier transform).[1]

A very general description of the main problems considered can be given as follows.

Suppose, we are considering a random field ξ in a region $T \subseteq R^d$ which is associated with a chaotic (stochastic) source η by means of the differential equation

$$(*) \qquad \mathcal{L}\xi = \eta$$

in T. A typical *chaotic* source can be represented by an appropriate random field η with independent values, i.e., generalized random function $\eta = (\varphi, \eta)$, $\varphi \in C_0^\infty(T)$, with independent random variables (φ, η) for any test functions φ with disjoint supports. The property of having independent values implies a certain "roughness" of the random field η which can only be treated functionally as a very irregular Schwarz distribution. With the lack of a proper development of nonlinear analyses for generalized functions, let us limit ourselves to the

[1] For related material see, for example, J.L. Lions, E. Magenes: *Nonhomogeneous Boundary Value Problems and Applications*, Springer, Berlin–Heidelberg–New York, 1972; L. Hörmander: *Analysis of Linear Partial Differential Operators*, Springer, Berlin–New York–Tokyo, 1983.

linear differential operator $\mathcal{L}$ where we can treat the equation (*) in a generalized sense. Assuming that its solution does exist, a variety of ξ is expected which satisfy the equation (*) with the given η. The problem is to find out what kind of additional information is required to determine ξ in the region T uniquely. One particular question is whether this kind of information can be given in the form of appropriate *boundary conditions*, related to a boundary behaviour of ξ on a boundary $\Gamma = \partial T$. Here the very question on the boundary behaviour is non-trivial, for ξ presumably can be described only as $\xi = (\varphi, \xi)$, $\varphi \in C_0^\infty(T)$, by means of the test functions φ with supports which can be set only in open neighborhoods of Γ. We are to think of the boundary behaviour of ξ in terms of events from the corresponding σ-algebra $\mathcal{A}(\Gamma) = \cap \mathcal{A}(\Gamma^\varepsilon)$ where σ-algebras $\mathcal{A}(\Gamma^\varepsilon)$ are generated by random variables (φ, ξ), $\operatorname{supp}\varphi \subseteq \Gamma^\varepsilon$, in Γ^ε-neighborhoods of Γ. A particular problem is to specify $\mathcal{A}(\Gamma)$ in appropriate terms for the ξ we are looking for. One way to do this is to define and specify for the generalized random function ξ the corresponding *boundary values* which generate $\mathcal{A}(\Gamma)$. Then, a certain type *boundary conditions*

$$(**) \qquad \xi|_\Gamma = \xi^+|_\Gamma$$

prescribing an appropriate family $\xi|_\Gamma$ of the boundary values by means of the given one as $\xi^+|_\Gamma$ can be set.

In general, the random source η in the region T and the family $\xi^+|_\Gamma$ of the given random variables on the boundary $\Gamma = \partial T$ are from certain functional classes. Accordingly, a solution of the boundary problem in the corresponding functional space $\mathbf{W}$ is required.

Note, in the case when ξ is expected to be chaotic like random fields with independent values, say, the corresponding σ-algebra $\mathcal{A}(\Gamma)$ is to be trivial, so there are no nontrivial boundary events to condition. As an illustration, one can take the known white noise $\xi = (\varphi, \xi)$, $\varphi \in L_2(T)$ on the interval $T = (0, 1)$ and consider the differential equation

$$\frac{d\xi}{dt} = \eta$$

which actually gives nothing more than $(\varphi, \xi) = -(\int \varphi(t)dt, \eta)$ over all $\varphi \in L_2(T)$, $\int_0^1 \varphi(t)dt = 0$; to determine ξ in a whole, additional information is required with respect to the nontrivial random variable $(1, \xi)$, independent on η, and this information cannot be supplied by means of any kind of "boundary conditions", since the σ-algebra $\mathcal{A}(\Gamma)$ is trivial and there are no nontrivial "boundary values". Thus, in this case there is no generalization of the integration by parts formula

$$(1,\xi) = \int_0^1 \xi(t)dt = [(1-c)\xi(1) + c\xi(0)] - \int_0^1 (t-c)\eta(t)dt$$

which could be used to supply the necessary additional information about ξ by a proper boundary value $[(1-c)\xi(1)+c\xi(0)] = \xi^+/\Gamma$ in an appropriate functional space $\mathbf{W} \ni \xi$. And in turn, when we have the functional class $\mathbf{W}$, the problem is, for the given differential equation (*), to find out all possible types of boundary conditions (**) so that arbitrary given these boundary conditions lead to the unique $\xi \in \mathbf{W}$.

As far as we get a model of the random field ξ, described in (*), (**) by stochastic PDE with the stochastic boundary conditions, we meet a variety of Probability/Stochastics problems related to the probability distribution of ξ and all its properties this distribution implies.

It happens that such well-known models as Levy Brownian Motion, Markov Free Field, Brownian Sheet, Ito Evolution Equation, etc., fit the framework suggested with the corresponding differential equation of elliptic, hyperbolic or parabolic type, and this greatly motivated me to develop the general model (*), (**) which could be of interest for Probability and Stochastic Physics.

At the very beginning,[2] my work on this project was greatly praised by A.N. Kolmogorov, and this was very stimulating. A few young followers were working with me at the time, and I would like to mention, in particular, V. Bulychev, V. Goryainov, N. Arato, and L. Markus. The main developments of the project on the whole were made thanks to my work at the Steklov Mathematical Institute, and though in different periods of time, there were fruitful visits to other institutions and I am grateful for their hospitality. In my acknowledgement I am especially grateful to S. Albeverio, A.N. Gushin, V.P. Mikhailov, L. Hörmander, O.A. Oleynick, and S.L. Sobolev for opportunities to present and discuss certain results on their seminars.

A complete revision of the book has been made after I started my work for CNR-IAMI (Milan) in March 1996.

To conclude, I wish to apologize for the lack of references and historical remarks – to make all of them properly would require writing

---

[2] See, for example, Yu.A. Rozanov: *Generalized Dirichlet Problem*, Dokl. AN USSR, **266**, N 6 (1982), 1067–1069.

another book. (Consequently, I tried to make the present one reasonably self-contained.)

Yu. A. Rozanov

Chapter I

Random Fields and Stochastic Sobolev Spaces

I.1. Generalized random fields

1.1. Introductionary concept

By randomness we assume the appropriate dependence on an elementary event $\omega \in \Omega$ of a probability space $(\Omega, \mathcal{A}, \mathrm{P})$ with a probability measure $\mathrm{P}(A)$, $A \subseteq \Omega$, on a σ-algebra of events $A \in \mathcal{A}$. In particular, random variables (real or complex valued) with the finite *mean*

$$\mathrm{E}\xi = \int_{\Omega} \xi(\omega)\, \mathrm{P}(d\omega)$$

are integrable functions $\xi = \xi(\omega)$, $\omega \in \Omega$. In what follows, we use the Hilbert space $\mathbf{H}$ of random variables ξ with the *meansquare norm*

$$\|\xi\| = (\mathrm{E}|\xi|^2)^{1/2}$$

where the scalar product

$$\langle \xi, \eta \rangle = \mathrm{E}\xi\bar{\eta}$$

is represented by the *covariance* of the random variables ξ, $\eta \in \mathbf{H}$. (Here and further on the overbar denotes complex conjugation.)

A term generalized *random field* we apply to a *generalized random function*

$$\xi = \xi(\varphi), \qquad \varphi \in D = C_0^\infty(T) \tag{1.1}$$

considered as linear continuous function in the Hilbert space **H** of random variables. Shortly, ξ is *meansquare linear continuous function* of $\varphi \in D$. Here $D = C_0^\infty(T)$ is the well known Schwartz space of infinitely differential functions $\varphi = \varphi(t)$, $t = (t_1, \ldots, t_d) \in T$, with compact supports $\operatorname{supp}\varphi \subseteq T$.

Recall the convergence $\varphi_n \to \varphi$ in the Schwartz space $D = C_0^\infty(T)$ means that for the given φ_n there is a compact in T where their supports are concentrated and all derivatives

$$\partial^k \varphi_n \to \partial^k \varphi,$$

$$\left(\partial^k = \frac{\partial^{|k|}}{\partial^{k_1} t_1 \ldots \partial^{k_d} t_d}, \quad k = (k_1, \ldots, k_d), \quad |k| = k_1 + \ldots + k_d\right)$$

converge uniformly. This is equivalent to the convergence of all derivatives with respect to the L_2-norm,

$$\|\partial^k \varphi_n - \partial^k \varphi\|_{L_2}^2 = \int |\partial^k \varphi_n - \partial^k \varphi|^2 \, dt \to 0,$$

since for all ψ with supports in the same compact we have

$$\|\psi\|_{L_2}^2 \le C \max |\psi|^2$$

and

$$\max |\psi|^2 \le C \|\psi'\|_{L_2}^2, \qquad \psi' = \frac{\partial^d \psi}{\partial t_1 \ldots \partial t_d}.$$

The known Hilbert space $L_2 = L_2(T)$ of measurable functions $f = f(t)$, $t \in T$, with the norm

$$\|f\|_{L_2} = \left(\int_T |f(t)|^2 \, dt\right)^{1/2}$$

will be often of our use. Note we have

$$L_2 = [D]$$

as closure of $D = C_0^\infty(T)$. It follows, say, from a fact that any bounded function f with a compact support in T can be approximated by *convolutions*

$$f * w = \int f(t - s) w(s) \, ds$$

with appropriate *delta-approximators* $w \in C_0^\infty(R^d)$ of a type

$$w \geq 0, \quad \int w(t)\,dt = 1, \qquad \int\limits_{|t|>\varepsilon} w(t)\,dt \to 0 \quad (\varepsilon > 0);$$

here we are to take $w \in C_0^\infty(O)$ in a sufficiently small neighborhood O of the origin, to get $f * w \in D = C_0^\infty(T)$.

Dealing with scalar generalized functions

$$f = (\varphi, f), \qquad \varphi \in D = C_0^\infty(T),$$

we denote their space (the Schwartz distributions space) as D^*. ■

Let $\xi = (\varphi, \xi)$, $\varphi \in C_0^\infty(T)$, be the generalized random function which is meansquare continuous with respect to a norm $\|\varphi\|_U$ in a Hilbert space U appeared as the closure $U = [C_0^\infty(T)]$ of the Schwartz test functions $\varphi \in C_0^\infty(T)$, that is

$$\|(\varphi, \xi)\| \leq C\|\varphi\|_U, \qquad \varphi \in C_0^\infty(T). \tag{1.2}$$

(Here and further on a symbol "C" denotes appropriate constants which may differ each from other.) For any $u \in U$ as a limit $u = \lim \varphi$ of $\varphi \in C_0^\infty(T)$ in the space $U = [C_0^\infty(T)]$ there is the corresponding meansquare limit

$$(u, \xi) = \lim_{\varphi \to u} (\varphi, \xi)$$

and it defines

$$\xi = (u, \xi), \qquad u \in U, \tag{1.3}$$

as the *generalized random function* on the Hilbert space U, we have in mind that ξ is meansquare linear continuous in it's dependence on $u \in U$. We consider $u \in U$ as the *test functions* for the random field ξ, assuming that the norm $\|\varphi\|_U$ in the Hilbert space $U = [C_0^\infty(T)]$ is continuous over the Schwartz space $D = C_0^\infty(T)$.

In this scheme the possibly largest test functions space can appear as the completeness $U_0 = [C_0^\infty(T)]$ of the Schwartz test functions space $D = C_0^\infty(T)$ with respect to the very meansquare norm

$$\|\varphi\| \stackrel{\text{def}}{=} (\mathrm{E}|(\varphi, \xi)|^2)^{1/2}.$$

This Hilbert space U_0 admits the embedding

$$U \subseteq U_0 \tag{1.4}$$

for any other test function space $U = [C_0^\infty(T)]$ thanks to the continuity condition (1.2). The scalar product in U_0,

$$\langle \varphi, u \rangle = \mathrm{E}(\varphi, \xi)\overline{(u, \xi)},$$

is given by the *covariance* of the random variables considered. As a function of $\varphi \in C_0^\infty(T)$, it determines for every $u \in U$ from the test functions space $U = [C_0^\infty(T)]$ an element of a dual space $U^* \subseteq D^*$, and it admits the known Riesz representation

$$\mathrm{E}(\varphi, \xi)\overline{(u, \xi)} = \langle \varphi, Ru \rangle_U \tag{1.5}$$

with the corresponding $Ru \in U$; here the linear continuous operator $R \geq 0$ is called *covariance operator* in the Hilbert space U.

In a case of the zero *mean*

$$\mathrm{E}\xi = (\varphi, \mathrm{E}\xi) \stackrel{\text{def}}{=} \mathrm{E}(\varphi, \xi), \qquad \varphi \in D = C_0^\infty(T),$$

which generally can be represented by any element $\mathrm{E}\xi \in U^* \subseteq D^*$, the formula (1.5) determines *correlation*, given by the correlation operator R in the Hilbert space U.

One can observe a variety of generalized random functions by the following construction. Let us take an arbitrary Hilbert space of the type $U = [C_0^\infty(T)]$ with a norm $\|\varphi\|_U$ continuous over the Schwartz test functions space $D = C_0^\infty(T)$. Let us take an arbitrary infinite-dimensional subspace $H \subseteq \mathbf{H}$ in the Hilbert space of random variables. Then any *unitary* mapping

$$\xi\colon U \ni u \to (u, \xi) \in H$$

defines the generalized random function $\xi = (u, \xi)$, $u \in U$, on the test functions space U. This construction for zero mean random variables gives us so-called *white noise ξ on the Hilbert space U*,

$$\mathrm{E}(u, \xi) = 0, \quad \mathrm{E}|(u, \xi)|^2 = \|u\|_U^2, \qquad u \in U.$$

1.2. Stochastic integral representations

We produce here a few constructions associated with appropriate stochastic integrals representing the generalized random functions on the test functions space

$$U = L_2(T) = [C_0^\infty(T)].$$

Let $\xi = \xi(t)$, $t \in T$, be a random function which is locally integrable as function in the Hilbert space $\mathbf{H}$ of random variables, having it's meansquare norm $\|\xi(t)\|$, $t \in T$, integrable over any bounded region $T_{\text{loc}} \subseteq T$,

$$\int_{T_{\text{loc}}} \|\xi(t)\| \, dt < \infty.$$

Then the integral

$$(\varphi, \xi) = \int \varphi(t)\xi(t) \, dt \tag{1.6}$$

determines the generalized random function $\xi = (\varphi, \xi)$, $\varphi \in D = C_0^\infty(T)$, with

$$\|(\varphi, \xi)\| \leq \max|\varphi| \cdot \int_{T_{\text{loc}}} \|\xi(t)\| \, dt$$

for all test functions having supports $\operatorname{supp}\varphi \subseteq T_{\text{loc}}$. According to

$$\|(\varphi, \xi)\| \leq \|\varphi\|_{L_2} \cdot \left(\int_T \|\xi(t)\|^2 \, dt \right)^{1/2},$$

stochastic integral (1.6) represents the generalized random function $\xi = (\varphi, \xi)$, $\varphi \in U = L_2(T)$, on the test functions space $L_2(T)$ when the condition

$$\int_T \|\xi(t)\|^2 \, dt < \infty$$

holds true.

Another type of *stochastic integral*

$$(\varphi, \xi) = \int \varphi(t) \, \eta(dt) \tag{1.7}$$

which represents the generalized random function $\xi = (\varphi, \xi)$, $\varphi \in U = L_2(T)$, can be defined with the appropriate *stochastic measure* $\eta(dt)$ in the Hilbert space $\mathbf{H}$ of random variables as aditive continuous function $\eta = \eta(B)$, $B \subseteq T$, on bounded Borel sets. For example, it can be the *stochastic measure with orthogonal* (*non-correlated*) *values* for disjoint sets, having the zero mean $\mathrm{E}\eta(dt) = 0$ and the corresponding *variance measure* $m(dt) = \mathrm{E}|\eta(dt)|^2$ such that

$$m(B) = \mathrm{E}|\eta(B)|^2 \leq C|B|, \qquad B \subseteq T,$$

where $|B|$ denotes the Lebesgue measure of B. Then for the stochastic integral (1.7) we have it's meansquare norm

$$\|(\varphi, \xi)\|^2 \leq C\|\varphi\|^2_{L_2(T)}.$$

Note that any generalized random function $\xi = (\varphi, \xi)$, $\varphi \in U = L_2(T)$, can be presented as stochastic integral of the type (1.7) by means of the corresponding stochastic measure $\eta(dt)$, which appears as

$$\eta(B) = (1_B, \xi)$$

with the indicators $\varphi = 1_B$ of bounded Borel sets $B \subseteq T$. Obviously, here we have the additive continuous function $\eta = \eta(B)$, $B \subseteq T$, in the Hilbert space $\mathbf{H}$ of random variables, since (φ, ξ) is meansquare linear continuous over all test functions $\varphi \in L_2(T)$. To define the very stochastic integral, we first set

$$\int \varphi(t)\, \eta(dt) = \sum_k x_k \eta(B_k)$$

for piece-wise constant functions of the type $\varphi = \sum_k x_k 1_{B_k}$ with values x_k on a finite number of disjoint sets $B_k \subseteq T$. Then, we obtain

$$\int \varphi(t)\, \eta(dt) = (\varphi, \xi).$$

In the space $U = L_2(T)$ any function $u \in U$ is a limit $u = \lim \varphi$ of the piece-wise constant functions, and there is the meansquare limit

$$\int u(t)\, \eta(dt) = \lim \int \varphi(t)\, \eta(dt)$$

which represents the corresponding random variables (u, ξ). ■

Let us consider a construction, similar to (1.6), in region of the direct product type $T = G \times I$. This construction is associated with a generalized random function $\xi_r = (x, \xi_r)$, $x \in C_0^\infty(G)$, in the region G, such that for each $x \in C_0^\infty(G)$ it represents the locally integrable random function (x, ξ_r), $r \in I$, in the region I with the meansquare norm

$$\|(x, \xi_r)\| \leq C_r \|x\|_{L_2(G)}$$

where C_r, $r \in I$, satisfies the condition

$$\int_I C_r^2\, dr < \infty.$$

We set

$$(\varphi, \xi) = \int_I (\varphi_r, \xi_r)\, dr \tag{1.8}$$

where $\varphi_r \in C_0^\infty(G)$ appear from the corresponding $\varphi \in C_0^\infty(T)$ in the region $T = G \times I$ with the variable $r \in I$ fixed. The very integral (1.8) can be defined first for functions of the type $\varphi = \sum_k x_k \delta_k$ with components $x_k \in C_0^\infty(G)$ and $\delta_k \in C_0^\infty(I)$, say. Taking in account that

$$\|\varphi\|_{L_2(T)} = \left(\int_I \|\varphi_r\|^2_{L_2(G)}\, dr \right)^{1/2},$$

we obtain the meansquare norm

$$\|(\varphi, \xi)\| \le \|\varphi\|_{L_2(T)} \cdot \left(\int_I C_r^2\, dr \right)^{1/2}.$$

In the space $U = L_2(T)$ any $u \in U$ is a limit $u = \lim \varphi$ of functions of the type suggested above, and there is a meansquare limit

$$\int_I (u_r, \xi_r)\, dr = \lim \int_I (\varphi_r, \xi_r)\, dr$$

which does define the stochastic integral (1.8) for all $u = \varphi \in C_0^\infty(T)$, in particular. Obviously, it represents the generalized random function $\xi = (u, \xi)$, $u \in U$, on the test functions space $U = L_2(T)$. ■

Finally, we consider a construction, similar to (1.7), associated with a *generalized stochastic measure with orthogonal* (*non-correlated*) *values* in the region $T = G \times I$. Let $\eta_B = (x, \eta_B)$, $x \in C_0^\infty(G)$, be a generalized random function in the region G, such that for every $x \in C_0^\infty(G)$ the corresponding (x, η_B), $B \subseteq I$, represents the stochastic measure in the region I with orthogonal (non-correlated) values (x, η_B) for all $x \in C_0^\infty(G)$ and disjoint sets B,

$$\mathrm{E}(x, \eta_B) = 0, \qquad \mathrm{E}|(x, \eta_B)|^2 \le C\|x\|^2_{L_2(G)}|B|$$

— recall, $|B|$ denotes the Lebesgue measure of $B \subseteq I$. We set

$$(\varphi, \xi) = \int_I \varphi_r\, \eta_{dr}. \tag{1.9}$$

The very *stochastic integral* in the right hand side is defined first for functions of the type $\varphi = \sum_k x_k 1_B$ with their functional values $x_k \in C_0^\infty(G)$ on a finite number of disjoint sets $B_k \subseteq I$, namely,

$$\int_I \varphi_r \, \eta_{dr} = \sum_k (x_k, \eta_{B_k}).$$

Then we have the meansquare norm

$$\|(\varphi, \xi)\| \leq C \|\varphi\|_{L_2(T)}$$

with

$$\|\varphi\|^2_{L_2(T)} = \sum_k \|x_k\|^2_{L_2(G)} |B_k|.$$

In the space $U = L_2(T)$ any function $u \in U$ is a limit $u = \lim \varphi$ of functions of the type considered above, and there is the meansquare limit

$$\int_I u_r \, \eta_{dr} = \lim \int_I \varphi_r \, \eta_{dr}$$

which does define the stochastic integral (1.9) for all $u = \varphi \in C_0^\infty(T)$, in particular. Obviously, it represents the generalized random function $\xi = (u, \xi)$, $u \in U$, on the test functions space $U = L_2(T)$.

1.3. Generalized random fields with independent values

The most chaotic random field seems to be considered as generalized random function with independent values in any disjoint neighborhoods.

We produce a construction representing random fields of this type. Let $\mu(dt)$ be a *Gaussian stochastic measure* with independent values in the region T,

$$\mathrm{E}\mu(dt) = 0, \qquad \mathrm{E}|\mu(dt)|^2 = F(dt)$$

having it's values $\mu(B)$ as the Gaussian random variables with the zero mean and the variance $\sigma^2 = F(B)$. Let $\nu(dr\,dt)$ be a centered *Poisson stochastic measure* with independent values in the direct product $R \times T$ of $R = R^1 \setminus \{0\}$ and the region T,

$$\mathrm{E}\nu(dr\,dt) = 0, \qquad \mathrm{E}|\nu(dr\,dt)|^2 = G(dr\,dt),$$

having it's values $\nu(\Delta \times B)$ as the centered Poisson random variables with the zero mean and the variance $\lambda = G(\Delta \times B)$. We form the generalized random function $\xi = (\varphi, \xi)$, $\varphi \in D = C_0^\infty(T)$, by means of the stochastic integral representation

$$(\varphi, \xi) = \int \varphi(t)\, \mu(dt) + \iint \varphi(t) r\, \nu(dr\, dt). \tag{1.10}$$

Obviously, it takes independent values (φ, ξ) for any test functions φ with disjoint supports,

$$\mathrm{E}(\varphi, \xi) = 0,$$

$$\mathrm{E}|(\varphi, \xi)|^2 = \int |\varphi(t)|^2 F\, dt + \iint |\varphi(t)|^2 r^2\, G(dr\, dt).$$

In particular, the random variables (φ, ξ) are meansquare continuous with respect to the L_2-norm $\|\varphi\|_{L_2}$ in a case of the absolute continuous measures

$$F(dt) = f(t)\, dt, \qquad \int\limits_R r^2\, G(dr\, dt) = g(t)\, dt \tag{1.11}$$

with bounded densities $f, g \leq C$.

Actually, we get here a general representation of random fields with independent values which are meansquare continuous over the test functions $\varphi \in D = C_0^\infty(T)$ with respect to the norm $\|\varphi\|_{L_2}$.

To explain it, we need a few prepositions proved. ■

Let us consider a *real* case of the generalized random function ξ with the mean $\mathrm{E}\xi = 0$ and independent values (φ, ξ) for the test functions $\varphi \in C_0^\infty(T)$ with disjoint supports. Suppose, it is meansquare continuous with respect to $\|\varphi\|_{L_2}$, so it can be treated as the generalized random function $\xi = (u, \xi)$, $u \in U$, on the test functions space $U = L_2(T)$. Then the corresponding stochastic measure $\eta(dt)$,

$$\eta(B) = (1_B, \xi), \qquad B \subseteq T,$$

is well defined.

LEMMA. *This is the stochastic measure with independent values.*

To prove it, we note first that in the L_2-space the indicator 1_B of any bounded closed set $B \subseteq T$ is the limit $1_B = \lim \varphi$ of functions $\varphi \in C_0^\infty(T)$ with supports $\operatorname{supp} \varphi \subseteq B^\varepsilon$ in an arbitrary small neighborhood B^ε of B, and correspondingly $\eta(B) = \lim(\varphi, \xi)$. It shows that the random variables $\eta(B)$ for this kind of disjoint sets B are independent. For any bounded Borel set B there are closed $B_n \subseteq B$ such that $1_B = \lim 1_{B_n}$, $\eta(B) = \lim \eta(B_n)$, and it shows that the random variables $\eta(B)$ are independent for disjoint B.

Considering this stochastic measure $\eta(dt)$ with independent values,

$$\mathrm{E}\eta(dt) = 0, \qquad \mathrm{E}|\eta(dt)|^2 = m(dt),$$

we have the variance measure

$$m(B) = \mathrm{E}|(1_B, \xi)|^2 \leq C|B|, \qquad B \subseteq T, \tag{1.12}$$

absolutely continuous with respect to the Lebesgue measure $|B|$ according to the continuity condition

$$\mathrm{E}|(\varphi, \xi)|^2 \leq C\|\varphi\|_{L_2}^2, \qquad \varphi \in L_2(T),$$

for any B the random variable. Obviously,

$$\eta(B) = \int_B \eta(dt)$$

has an infinite divisible probability distribution, since it can be represented as a sum of series of infinitely small independent random variables. And, according to the known Levy–Khinchine formula, it's characteristic function is

$$\log \mathrm{E}e^{i\lambda\eta(B)} = -\frac{1}{2}\lambda^2 F(B) +$$

$$+ \int_R (e^{i\lambda r} - 1 - i\lambda r)\, G(dr \times B), \qquad -\infty < \lambda < \infty,$$

with the corresponding parameters F and G which depend on $B \subseteq T$ in such a way that $F(B)$ and $G(dr \times B)$ represent some measures, since $\eta(B)$ is stochastic measure with independent values,

$$m(B) = \mathrm{E}|\eta(B)|^2 = F(B) + \int_R r^2\, G(dr \times B).$$

Hence, for linear combinations $\varphi = \sum \lambda_k 1_{B_k}$ with disjoint $B_k \subseteq T$ and

$$(\varphi, \xi) = \sum_k \lambda_k \eta(B_k)$$

we get

$$\log \mathrm{E}e^{i(\varphi,\xi)} = -\frac{1}{2}\sum_k \lambda_k^2 F(B_k) + \sum_k \int_R (e^{i\lambda_k r} - 1 - i\lambda_k r)\, G(dr \times B_k) =$$

$$= -\frac{1}{2}\int_T |\varphi(t)|^2\, F(dt) + \int_T\int_R [e^{i\varphi(t)r} - 1 - i\varphi(t)r]\, G(dr\, dt)$$

where we can go to any $\varphi \in C_0^\infty(T)$ by the obvious limit. This leads us to the *characteristical functional* of the generalized random field $\xi = (\varphi, \xi)$, $\varphi \in C_0^\infty(T)$, which is exactly the same as in a case of the stochastic integral representation (1.10) when we have

$$\log \mathrm{E}e^{i(\varphi,\xi)} = -\frac{1}{2}\int |\varphi(t)|^2\, F(dt) +$$

$$+ \iint [e^{i\varphi(t)r} - 1 - i\varphi(t)r]\, G(dr\, dt), \qquad \varphi \in C_0^\infty(T). \qquad (1.13)$$

Having in mind to make no difference for ξ with the same probabilities distributions, we can conclude with the following preposition.

THEOREM. *The generalized random field ξ with independent values, meansquare continuous with respect to $\|\varphi\|_{L_2}$ over the test functions $\varphi \in C_0^\infty(T)$, can be considered in a form of the Levy–Khinchine–Ito representation* (1.10).

Remind, that here the conditions (1.11), (1.12) must be taken in account./ In particular, we can conclude that the property to be with independent values involves a certain roughness of the random field $\xi = (\varphi, \xi)$, $\varphi \in C_0^\infty(T)$, which cannot be meansquare continuous with respect to a weaker norm than the weighted L_2-norm $\|\varphi\|^2 = \mathrm{E}|(\varphi, \xi)|^2$ appeared from (1.10).

Of course, there are random fields with independent values which are quite different from the type (1.10). For example, taking the Dirac delta-functions δ_{t_k} in various points $t_k \in T$ and independent random variables ξ_k, $k = 1, \dots, n$, we can set on T a random field

$$\xi = \sum_k \xi_k \delta_{t_k}$$

with the independent values

$$(\varphi, \xi) = \sum_k \xi_k \varphi(t_k), \qquad \varphi \in C_0^\infty(T),$$

for the test functions φ having disjoint supports.

I.2. Random fields realizations

2.1. Some embeddings

Let $\xi = (\varphi, \xi)$, $\varphi \in D = C_0^\infty(T)$, be a generalized random function. It's values (φ, ξ) as random variables depend on an outcome $\omega \in \Omega$ which gives the corresponding *realization*

$$\xi_\omega = (\varphi, \xi_\omega), \qquad \varphi \in D = C_0^\infty(T),$$

called also the *trajectory*, or the *sample function*, of the random ξ. Note, that with no difference in the values $(\varphi, \xi) \in \mathbf{H}$ of the random function ξ in the Hilbert space $\mathbf{H}$ of random variables, as we agree to treat it — see (1.1), we can arbitrary change every (φ, ξ_ω) correspondingly on any ω-set of the zero probability, and this gives a *modification* of the random function ξ which all realizations ξ_ω can be radically different from the former ones.

We are interested to deal with a modification of the generalized random function ξ such that all possible realizations ξ_ω are generalized functions (Schwartz distributions) from an appropriate functional space X which can be characterized as the dual space $X = U^*$ to some Hilbert space of the type $U = [C_0^\infty(T)]$ serving as the test functions space for ξ, that is

$$\xi_\omega \in X = U^* \subseteq D^*, \qquad \omega \in \Omega. \tag{2.1}$$

To choose the test functions space U, we are to take in account the embedding

$$U \subseteq U_0 \tag{2.2}$$

to the proper U_0 — see (1.4).

From now we assume that $U_0 = [C_0^\infty(T)]$ is some given test functions space for ξ (not necessary the former one). We call (2.2) the *Hilbert–Schmidt embedding*, if for an orthonormal basis $\{\varphi_k\}$ in the Hilbert space U we have

$$\sum_k \|\varphi_k\|_{U_0}^2 < \infty \tag{2.3}$$

in the Hilbert space U_0.

THEOREM. *In a case of* (2.2) *being the Hilbert–Schmidt embedding, there is the modification of ξ with all realizations ξ_ω of the type* (2.1).

PROOF. Let us start from the condition (2.3) and take $\xi_k = (\varphi_k, \xi)$. For the random variables $\xi_k = \xi_k(\omega)$, $\omega \in \Omega$, we have

$$\mathrm{E}\sum_k |\xi_k(\omega)|^2 = \sum_k \mathrm{E}|(\varphi_k, \xi)|^2 \leq C \sum_k \|\varphi_k\|^2_{U_0} < \infty. \tag{2.4}$$

Hence

$$\sum_k |\xi_k(\omega)|^2 < \infty$$

for almost all $\omega \in \Omega$. Let us change the values $\xi_k = (\varphi_k, \xi)$ in a way, that the series above converges for all $\omega \in \Omega$. Let us set

$$\tilde{\xi}_\omega = \sum_k \xi_k(\omega) x_k \in X$$

with the dual orthonormal basis $\{x_k\}$ in the dual space $X = U^*$,

$$(\varphi_j, x_k) = \begin{cases} 1, & k = j, \\ 0, & k \neq j. \end{cases}$$

For any $\omega \in \Omega$ we have

$$\tilde{\xi}_\omega = (\varphi, \tilde{\xi}_\omega), \qquad \varphi \in D \subseteq C_0^\infty(T)$$

as realization of the generalized random function $\tilde{\xi}$ with values

$$(\varphi, \tilde{\xi}) = \sum_k \tilde{\xi}_k(\varphi, x_k) = \sum_k (\varphi, x_k)(\varphi_k, \xi) =$$

$$\left(\sum_k (\varphi, x_k)\varphi_k, \xi\right) = (\varphi, \xi)$$

in the Hilbert space $\mathbf{H}$ of random variables, since

$$\varphi = \sum_k (\varphi, x_k)\varphi_k = \lim_{n\to\infty} \sum_{k=1}^n (\varphi, x_k)\varphi_k$$

in the Hilbert space U and the values (φ, ξ) are meansquare continuous with respect to convergence in U. Having $\tilde{\xi}$ as the modification of ξ, we see that all realizations $\tilde{\xi}_\omega$ are elements of the dual space $X = U^*$. The theorem is proved. ■

We essentially used above the condition (2.4) which can be expressed in terms of the covariance operator R in the Hilbert space U as

$$\sum_k \langle \varphi_k, R\varphi_k \rangle_U < \infty \tag{2.5}$$

— see (1.5). This means that the covariance operator R is *nuclear*. Thus, we have the following result with respect to the realizations of the generalized random function $\xi = (u, \xi)$, $u \in U$, on the test functions space $U = [C_0^\infty(T)]$.

THEOREM. *In a case of the nuclear covariance operator in the Hilbert space U all realizations ξ_ω are of the type* (2.1).

Note, for any test functions space U_0 the *Hilbert–Schmidt* embedding (2.2) is equivalent to the same type embedding

$$X_0 \subseteq X \tag{2.2$'$}$$

for the dual space $X_0 = U_0^*$.

Indeed, taking an arbitrary orthonormal basis $\{x_j\}$ in the Hilbert space $X_0 = U_0^*$, we can express the condition (2.3) in the form

$$\sum_{k,j} |(\varphi_k, x_j)|^2 < \infty, \tag{2.3$'$}$$

since for the dual orthonormal basis $\{u_j\}$ in U_0 we have

$$\|\varphi\|_{U_0}^2 = \sum_j |\langle \varphi, u_j \rangle_{U_0}|^2 = \sum_j |(\varphi, x_j)|^2$$

for any $\varphi \in U_0$ and

$$\sum_k \|\varphi_k\|_{U_0}^2 = \sum_{k,j} |(\varphi_k, x_j)|^2.$$

Thus, along with the embedding $U \subseteq U_0$, we have the dual embedding

$$X_0 = U_0^* \subseteq U^* = X.$$

In its turn, $U = X^*$ is dual to the Hilbert space $X = U^*$ and $\{\varphi_k\}$ is the orthogonal basis in X^*. Accordingly, the condition (2.3)$'$ means also that $U_0^* \subseteq U^*$ is the Hilbert–Schmidt embedding.

2.2. Gaussian random fields

A real generalized random function ξ is called *Gaussian*, if all it's values (φ, ξ), $\varphi \in C_0^\infty(T)$, are Gaussian random variables.

Let us consider the Gaussian ξ with the zero mean $\mathrm{E}\xi = 0$ on some Hilbert test functions space $U = [C_0^\infty(T)]$. The characteristical functional of ξ is described by the formula

$$\log \mathrm{E} e^{i(\varphi, \xi)} = -\frac{1}{2} \langle \varphi, R\varphi \rangle_U$$

with the corresponding covariance (correlation) operator R. Recall, in a case of the nuclear R there is the modification of ξ with all realizations of the type (2.1) — see (2.5). For the Gaussian ξ the inverse statement holds true as follows.

THEOREM. *If with a positive probability*

$$\xi_\omega \in X = U^*$$

then the correlation operator R is to be nuclear.

To prove it, we need two lemmas on general Gaussian variables ξ_k with the zero mean $\mathrm{E}\xi_k = 0$ and the variance $\sigma_k^2 = \mathrm{E}\xi_k^2$, $k = 1, 2, \ldots$.

LEMMA 1. *If* $\lim \xi_k = 0$ *with a positive probability, then* $\lim \sigma_k^2 = 0$.

Indeed, with

$$A: \{\omega: \lim \xi_k(\omega) = 0\}, \qquad \mathrm{P}(A) = p > 0$$

in a case of $\overline{\lim}\sigma_k^2 \neq 0$ we have

$$\overline{\lim}\mathrm{P}(|\xi_k| > \varepsilon) \geq 1 - p/2$$

for a sufficiently small $\varepsilon > 0$, since

$$\mathrm{P}(|\xi_k| > \varepsilon) = \int\limits_{|x|>\varepsilon} \frac{1}{\sqrt{2\pi}\sigma_k} e^{-\frac{x^2}{2\sigma_k^2}}\, dx.$$

Therefore it should be

$$\overline{\lim}\mathrm{P}(\{|\xi_k| > \varepsilon\} \cap A) \geq p/2,$$

and this contradicts to the convergence $\xi_k(\omega) \to 0$ for $\omega \in A$.

LEMMA 2. *If* $\sum \xi_k^2 < \infty$ *with a positive probability for independent* ξ_k, $k = 1, 2, \ldots$, *then*

$$\mathrm{E}\sum \xi_k^2 = \sum \sigma_k^2 < \infty.$$

Indeed, according to Lemma 1, $\sigma_k^2 \to 0$ and

$$\delta_k^2 = \int\limits_{|x|>1} x^2 \frac{1}{\sqrt{2\pi}\sigma_k} e^{-\frac{x^2}{2\sigma_k^2}}\, dx = o(\sigma_k^2).$$

Therefore, the independent $\hat{\xi}_k = \xi_k 1_{\{|\xi_k| \leq 1\}}$, $k = 1, 2, \ldots$, have

$$\mathrm{E}\hat{\xi}_k^2 = \sigma_k^2 - \delta_k^2 \sim \sigma_k^2$$

and, according to the known "zero-one low" and "theorem on three series," the convergence $\sum_k \hat{\xi}_k^2 < \infty$ implies the convergence

$$\sum \mathrm{E}\hat{\xi}_k^2 \sim \sum \sigma_k^2 < \infty.$$

Now we are able to prove the very theorem. Having with a positive probability $\xi_\omega \in X$ in the dual Hilbert space $X = U^*$ we get

$$\xi_k(\omega) = (\varphi_k, \xi_\omega) \to 0, \qquad \sum \xi_k(\omega)^2 = \|\xi_\omega\|_X^2$$

for an orthonormal basis $\{\varphi_k\}$ in the Hilbert space U. According to Lemma 1,

$$\mathrm{E}\xi_k^2 = \langle \varphi_k, R\varphi_k \rangle_U \to 0.$$

Therefore R is the *compact* operator. For the orthonormal basis $\{\varphi_k\}$, formed with it's eigen functions, we have independent Gaussian variables $\xi_k = (\varphi_k, \xi)$, $k = 1, 2, \ldots$, since

$$\mathrm{E}\xi_k\xi_j = \langle \varphi_k, R\varphi_j \rangle_U = \begin{cases} \sigma_k^2, & j = k, \\ 0, & j \neq k. \end{cases}$$

According to Lemma 2,

$$\sum \sigma_k^2 < \infty$$

and we see that the operator R is nuclear. The theorem is proved.

I.3. Test functions spaces

3.1. Spaces of the W-type

In what will follow later on, we apply test functions spaces $W = [D]$ which appear as completion of $D = C_0^\infty(T)$ with respect to scalar product of the type

$$\langle \varphi, u \rangle_W = (\varphi, \mathcal{P}\bar{u}), \qquad \varphi, u \in D, \tag{3.1}$$

defined by means of a symmetric differential operator

$$\mathcal{P} = \sum a_k \partial^k \geq 0$$

with *real* infinitely differential coefficients,

$$\mathcal{P}\varphi = \mathcal{P}^*\varphi = \sum (-1)^{|k|} \partial^k (a_k \varphi), \qquad (\varphi, \mathcal{P}\bar{\varphi}) \geq 0, \qquad \varphi \in D.$$

Considering the formula (3.1) in it's dependence on $\varphi \in D$, we have the known Riesz formula which represents the corresponding element

$$\mathcal{P}\bar{u} = \overline{\mathcal{P}u} \in X = W^*$$

from the dual space $X = W^*$. Obviously, this very element $\mathcal{P}\bar{u} \in X$, considered just as linear continuous function of $\varphi \in D \subseteq W$, can be identified with the corresponding Schwartz distribution $\mathcal{P}\bar{u} \in D^*$, since D is dense in the Hilbert space $W = [D]$ and the norm

$$\|\varphi\|_W = (\varphi, \mathcal{P}\bar{\varphi})^{1/2} = (\bar{\varphi}, \mathcal{P}\varphi)^{1/2} = \|\bar{\varphi}\|_W$$

is continuous over the Schwartz test functions space $D = C_0^\infty(T)$.

The operator

$$\mathcal{P}\colon\ D \to X = W^*$$

is *isometric* on the Hilbert space W, that is

$$\|\mathcal{P}\bar{u}\|_X = \sup_{\|\varphi\|_W \le 1} (\varphi, \mathcal{P}\bar{u}) =$$

$$= \sup_{\|\varphi\|_W \le 1} \langle\varphi, u\rangle_W = \|u\|_W = \|\bar{u}\|_W, \qquad u \in D,$$

and it's extension on the whole space $W = [D]$ gives the unitary operator $\mathcal{P}$ in the Riesz representation

$$x = \mathcal{P}\bar{u} = \langle\varphi, u\rangle, \qquad \varphi \in W,$$

of any $x \in X = W^*$ with the corresponding $u \in W$,

$$\mathcal{P}W = X = W^*.$$

Taking in account (3.1) with our differential operator $\mathcal{P}$ as the Riesz formula, we see that $\mathcal{P}D$ is dense in the dual Hilbert space $X = W^*$, for D is dense in the very W. Thus, we are getting

$$X = [\mathcal{P}D]$$

as the closure of $\mathcal{P}D$.

Suppose, the differential operator $\mathcal{P}$ satisfies *semiboundness* condition of the form

$$\|u\|_W = (u, \mathcal{P}u)^{1/2} \ge c \int_{T_{\text{loc}}} |u(t)|\, dt, \qquad u \in D, \tag{3.2}$$

in any bounded region T_{loc} with the closure $[T_{\text{loc}}] \subseteq T$. (For illustration, one can take

$$\mathcal{P} = (-1)^d \frac{\partial^{2d}}{\partial t_1^2 \ldots \partial t_d^2},$$

say.) Then, having

$$|(u, \varphi)| \leq \max |\varphi| \int\limits_{T_{\text{loc}}} |u(t)|\, dt \leq C \max |\varphi| \cdot \|u\|_W, \qquad u \in D,$$

for $\varphi \in D$ with supports $\operatorname{supp} \varphi \subseteq T_{\text{loc}}$, we see that all $\varphi \in D$ as the Schwartz distributions $\varphi = (u, \varphi)$, $u \in D$, are continuous with respect to the norm $\|u\|_W$. Therefore, $D \subseteq X = W^*$, and actually we have

$$X = [D]$$

since $X = [\mathcal{P}D]$, $\mathcal{P}D \subseteq D$. We see also that

$$\|\varphi\|_X = \sup_{\|u\|_W \leq 1} |(u, \varphi)| \leq C \max |\varphi|$$

for $\varphi \in D$ with $\operatorname{supp} \varphi \subseteq T_{\text{loc}}$, so the norm $\|\varphi\|_X$ is continuous over the Schwartz space $D = C_0^\infty(T)$. Thus, we are getting $X = W^*$ as what we will call a *space of the W-type*, having in mind that

$$X = [D] \subseteq D^*$$

with the norm $\|\varphi\|_X$ continuous over the Schwartz test functions space $D = C_0^\infty(T)$, similar to what we have with respect to the former W,

$$W = [D] \subseteq D^*$$

being in it's turn the dual space to $X = [D]$.

The differential operator $\mathcal{P}$ is well defined on the Schwartz distributions $u \in D^*$ as

$$\mathcal{P}u = (\varphi, \mathcal{P}u) = (\mathcal{P}^*\varphi, u), \qquad \varphi \in D,$$

and in this way it represents the *unitary* extension

$$\mathcal{P}\colon W \to X = \mathcal{P}W = W^* \tag{3.3}$$

of the *isometric* operator $\mathcal{P}$ on $D \subseteq W$ in the Hilbert space $W = [D]$ which has been considered above in association with the Riesz formula. Indeed, we have

$$(\mathcal{P}^*\varphi, \bar{u}) = (\varphi, \mathcal{P}\bar{u}) = \langle\varphi, u\rangle_W, \qquad \varphi \in D,$$

for all elements $u \in D$ and, consequently, for all its limits $u \in W$ in the Hilbert space $W = [D]$.

We conclude here with the dual Riesz formula

$$\langle\varphi, u\rangle_W = \langle\mathcal{P}\varphi, \mathcal{P}u\rangle_X = (\varphi, \mathcal{P}\bar{u}) \tag{3.4}$$

for all $\varphi, u \in W$. ■

As an example, let us take the differential operator

$$\mathcal{P} = \sum_{|k|\leq p} (-1)^{|k|}\partial^{2k}$$

defining the scalar product (3.1) as

$$\langle u, v\rangle_W = (u, \mathcal{P}\bar{v}) = \sum_{|k|\leq p} \langle\partial^k u, \partial^k v\rangle_{L_2}, \qquad u, v \in D.$$

With this we are getting the well-known *Sobolev spaces*

$$W = \overset{\circ}{W}{}_2^p(T), \qquad X = \mathcal{P}W = W^* = \overset{\circ}{W}{}_2^{-p}(T);$$

the norm there we denote by

$$\|u\|_W = \|u\|_p, \qquad \|x\|_X = \|x\|_{-p}.$$

Recall,

$$\|u\|_p^2 = \sum_{|k|\leq p} \|\partial^k u\|_{L_2}^2$$

not only for $u \in D = C_0^\infty(T)$ but also for all $u \in \overset{\circ}{W}{}_2^p(T)$ with $\partial^k u$ as the generalized derivatives,

$$\partial^k u = (\varphi, \partial^k u) = (-1)^{|k|}(\partial^k\varphi, u), \qquad \varphi \in D.$$

In particular, here we assert that for all $u \in \overset{\circ}{W}{}_2^p(T)$ *generalized derivatives* $\partial^k u$, $|k| \leq p$, *belong to the space* $L_2 = L_2(T)$. One can see it as follows. Any function $u = \overset{\circ}{W}{}_2^p(T)$ is the limit $u = \lim u_n$

of functions $u_n \in D$ in $\overset{\circ}{W}{}^2_p(T) = [D]$. For such functions the limit $\lim \partial^k u_n = f_k$ exists in L_2 for all $\varphi \in D$,

$$(\varphi, \partial^k u) = (-1)^k(\partial^k \varphi, u) = \lim(-1)^k(\partial^k \varphi, u_n)$$
$$= \lim(\varphi, \partial^k u_n) = (\varphi, f_k),$$

so $\partial^k u = f_k \in L_2$ and

$$\|u\|_p^2 = \lim \|u_n\|_p^2 = \lim \sum_{|k|\le p} \|\partial^k u_n\|_{L_2}^2 = \sum_{|k|\le p} \|f_k\|_{L_2}^2.$$

An analogous example can be given by the differential operator

$$\mathcal{P} = \sum_{j=1}^{m} \sum_{|k_j|\le p_j} (-1)^{k_j} \partial^{2k_j}$$

defining the scalar product (3.1) as

$$\langle u, v\rangle_W = (u, \mathcal{P}\bar{u}) = \sum_{j=1}^{m} \sum_{|k_j|\le p_j} \langle \partial^{k_j} u, \partial^{k_j} v\rangle_{L_2}, \qquad u, v \in D,$$

where the multi-indices k_j denote derivatives in the jth group of variables, formally united into multi-dimensional $t_j \in R^{d_j}$ and, for $\sum_{j=1}^m d_j = d$, representing $t = (t_1, \ldots, t_m) \in R^d$. The corresponding

$$W = \overset{\circ}{W}{}^p_2(T), \qquad X = \mathcal{P}W = W^* = \overset{\circ}{W}{}^{-p}_2(T)$$

with the multi-index $p = (p_1, \ldots, p_m)$ will be called *anisotropic Sobolev spaces* with the same notation employed for

$$\|u\|_W = \|u\|_p, \qquad \|x\|_X = \|x\|_{-p}.$$

Clearly, for all $u \in \overset{\circ}{W}{}^2_p(T)$, the generalized derivatives $\partial^{k_j} u$, $k_j \le p_j$, belong to the space $L_2 = L_2(T)$ and

$$\|u\|_p^2 = \sum_{j=1}^{m} \sum_{|k_j|\le p_j} \|\partial^{k_j} u\|_2^2, \qquad u \in \overset{\circ}{W}{}^2_p(T).$$

Obviously, we have the embedding

$$\overset{\circ}{W}{}^p_2(T) \subseteq L_2(T) \subseteq \overset{\circ}{W}{}^{-p}_2(T)$$

where $L_2(T)$ represents in our scheme (3.1)–(3.4) the space $W = X = \mathcal{P}W = W^*$ with the identity operator $\mathcal{P} = I$. ■

Let us turn to the arbitrary $W = [D]$ and corresponding $X = W^*$. Let us assume that there exists a *non-degenerate embedding*

$$W \subseteq W_0 \tag{3.5}$$

into the Hilbert space $W_0 = [D]$ with $X_0 = W_0^*$ of the W type; here the condition for a weaker norm $\|u\|_{W_0} \leq C\|u\|_W$, $u \in W$, to be non-degenerate means that $\|u\|_{W_0} \neq 0$ if $\|u\|_W \neq 0$.

THEOREM. *The space X is also of the W-type with the dense embeddings*

$$D \subseteq X_0 = W_0^* \subseteq W^* = X.$$

PROOF. Obviously, any element $x \in X_0$ determines the linear continuous functional $x = (u, x)$ on $u \in W \subseteq W_0$ and accordingly $X_0 \subseteq X$ with

$$\|x\|_X = \sup_{\|u\|_W \leq 1} |(u,x)| \leq C \sup_{\|u\|_{W_0} \leq 1} |(u,x)| \leq C\|x\|_{X_0}.$$

Moreover, X_0 is dense in X, otherwise there should exist an element $u \in W = X^*$ not equal to zero such that $(x, u) = 0$ for all $x \in X_0$, determining the zero element $u \in W_0 = X_0^*$, which contradicts the condition $\|u\|_{W_0} \neq 0$ for $\|\bar{u}\|_W \neq 0$.

In the given $X_0 = [D]$ any element $x_0 \in X_0$ can be approximated by $\varphi \in D$, and accordingly, with respect to the weaker norm in $X \supseteq X_0$ we have

$$\|x_0 - \varphi\|_X \leq C\|x_0 - \varphi\|_{X_0} \to 0.$$

The dense embeddings $D \subseteq X_0$, $X_0 \subseteq X$ in X give $X = [D]$ with the norm $\|\varphi\|_X$, $\varphi \in D$, continuous for the convergence in $D = C_0^\infty(T)$ since the stronger norm $\|\varphi\|_{X_0}$ is with this property.

Actually, we have a nondegenerate embedding $X_0 = W_0^* \subseteq W^* = X$; indeed, for $x \in X_0$ with $\|x\|_X = 0$ we have $(u, x) = 0$, $u \in W$ and since $W = [D]$ is dense in W_0 we have $(u, x) = 0$ for all $u \in W_0$, that is, $x = 0$ as element of the dual space $X_0 = W_0^*$, $\|x\|_{X_0} = 0$.

For example, as a sample of the W-type space in the criterion (3.5), we can apply any Sobolev space $W_0 = \overset{\circ}{W}{}_2^{\pm p}(T)$ to compare it with the other space W, we are interested in. ■

We are to stress that in the dual W-type spaces

$$W = [D] = X^* \subseteq D^*, \qquad X = [D] = W^* \subseteq D^*$$

the corresponding bilinear form

$$(u, x) \equiv (x, u), \qquad u \in W = [D], \qquad x \in X = [D]$$

identifies elements $u \in W = X^*$ and $x \in X = W^*$ as the certain Schwartz distributions

$$u = (x, u) \equiv (u, x), \qquad x \in D,$$

and

$$x = (u, x) \equiv (x, u), \qquad u \in D = C_0^\infty(T).$$

3.2. The Fourier transform of W-spaces

We briefly recall some properties of the *Fourier transform*

$$\tilde{\varphi}(\lambda) = \frac{1}{(2\pi)^{d/2}} \int_{R^d} e^{-i\lambda t} \varphi(t)\, dt, \qquad \lambda \in R^d.$$

For integrable $\varphi = \varphi(t)$, $t \in R^d$, this formula determines a continuous function of the variable λ, $\tilde{\varphi}(\lambda) \to 0$ as $\lambda \to \infty$, and in the case of integrable $\tilde{\varphi}(\lambda)$, $\lambda \in R^d$, we get

$$\varphi(t) = \frac{1}{(2\pi)^{d/2}} \int_{R^d} e^{i\lambda t} \tilde{\varphi}(\lambda)\, d\lambda, \qquad t \in R^d,$$

by the *reverse Fourier transform.* Assuming integrability of $|t|^p \varphi(t)$, $t \in R^d$, with differentiation under the integral, we obtain

$$\partial^k \tilde{\varphi} = \widetilde{(-it)^k \varphi}, \qquad |k| \leq p.$$

On the other hand, if the very function φ has integrable derivatives $\partial^k \varphi$, $|k| \leq p$, then its Fourier transform with integration by parts gives

$$\widetilde{\partial^k \varphi} = (i\lambda)^k \tilde{\varphi}.$$

For the convolution $u * v$ it's Fourier transform with iterated integration gives

$$\widetilde{u * v} = (2\pi)^{d/2} \tilde{u} \tilde{v}.$$

Taking here $v(-t)$, $t \in R^d$, instead of v, by the reverse Fourier transform we obtain

$$\int e^{i\lambda s}\tilde{u}(\lambda)\overline{\tilde{v}(\lambda)}\,d\lambda = \int u(s+t)\overline{v(t)}\,dt,$$

and it leads to the *Parseval equality*

$$\int \tilde{u}(\lambda)\overline{\tilde{v}(\lambda)}\,d\lambda = \int u(t)\overline{v(t)}\,dt.$$

Let us turn to the known space J of infinitely differentiable functions φ with their derivatives $\partial^k\varphi(t)$ rapidly decreasing at infinity in a way that

$$|t|^p\partial^k\varphi(t) \to 0, \qquad t \to \infty,$$

for any $p \geq 0$. Convergence $\varphi_n \to \varphi$ in J is defined as uniform convergence

$$|t|^p\partial^k\varphi_n \to |t|^p\partial^k\varphi$$

for all derivatives $\partial^k\varphi$ and $p = 0, 1, \ldots$, say. There is the obvious embedding

$$D = C_0^\infty(T) \subseteq J$$

with convergence $\varphi_n \to \varphi$ in D stronger than in J. The Fourier transform of a function $\varphi \in J$ gives $\tilde{\varphi} \in \tilde{J}$ from the space $\tilde{J} = J$ of the same type with functions of the dual variable $\lambda \in R^d$, and the same property holds for the reverse Fourier transform, so we have the bijective correspondence

$$J \ni \varphi \leftrightarrow \tilde{\varphi} \in \tilde{J}.$$

This correspondence is continuous, that is for $\varphi_n \to \varphi$ in J we have $\tilde{\varphi}_n \to \tilde{\varphi}$ in $\tilde{J}$ and vice versa.

Taking $w(t) \in C_0^\infty(R^d)$ with $w(t) = 1$ for $|t| \leq 1$ and putting $\varphi_n(t) = w(t/n)\varphi(t)$ for any function $\varphi \in J$, we obtain $\varphi_n \in D = C_0^\infty(R^d)$ such that $\varphi_n \to \varphi$, $\tilde{\varphi}_n \to \tilde{\varphi}$ in J, so $D = C_0^\infty(T)$ is dense in J and the Fourier transform $\tilde{D}$ is dense in $\tilde{J} = J$ as well. Every Schwartz distribution $u \in D^*$,

$$u = (\varphi, u), \qquad \varphi \in D,$$

which is continuous with respect to convergence in J, can be uniquely extended by continuity on all $\varphi \in J$, and this defines the *generalized function* $u \in J^* \subseteq D^*$.

For any generalized function $u \in J^*$, the equality

$$(\varphi, \tilde{u}) = (\tilde{\varphi}, u)$$

defines, with the help of $(\tilde{\varphi}, u)$ on $\tilde{\varphi} \in J$, the generalized function $\tilde{u} \in J^*$ on $\varphi \in J$ which is the generalized Fourier transform of $u \in J^*$. (In a similar way the *reverse generalized Fourier transform* is defined.) For example,

$$\frac{1}{(2\pi)^{d/2}} \int e^{-i\lambda t} \varphi(\lambda)\, d\lambda = \tilde{\varphi}(t) = (\tilde{\varphi}, \delta_t) = (\varphi, \tilde{\delta}_t), \qquad \varphi \in J,$$

defines the function $\tilde{u} = \delta_t(\lambda) = \frac{1}{(2\pi)^{d/2}} e^{-i\lambda t}$, $\lambda \in R^d$, as the generalized Fourier transform of the delta function δ_t at the point $t \in R^d$.

The Fourier transform of functions $u \in L_2(R^d)$ represents the unitary mapping

$$L_2(R^d) \leftrightarrow L_2(R^d)$$

thanks to the Parseval equality. ■

Let us consider the Hilbert space of the type $W = [D]$ with $D = C_0^\infty(T)$ in $T = R^d$ which appears as completion of D with respect to the scalar product

$$\langle \varphi, u \rangle_W = \int \tilde{\varphi}(\lambda) \overline{\tilde{u}(\lambda)} F(\lambda)\, d\lambda, \qquad \varphi, u \in D, \tag{3.6}$$

given by means of the Fourier transform and a *weight function* F,

$$0 \le F(\lambda) \le C(1 + |\lambda|^{2p}), \qquad \lambda \in R^d.$$

We associate this $W = [D]$ with the corresponding Hilbert space $L_{2,F}$ where the scalar product is given as

$$\langle f, g \rangle_{L_{2,F}} = \int f(\lambda) \overline{g(\lambda)} F(\lambda)\, d\lambda,$$

and the very $L_{2,F}$ consists of the functions f,

$$f F^{1/2} \subset L_2(R^d).$$

Note, that $D = C_0^\infty(R^d)$ is dense in the space $L_{2,F}$ (for example, any bounded function f with compact support can be approximated by convolutions $f * w$ with appropriate delta-approximators). Therefore, the space $J \supseteq D$ of the rapidly decreasing functions is dense in $L_{2,F}$,

$$L_{2,F} = [J].$$

Obviously, the isometric correspondence

$$W \ni \varphi \leftrightarrow \tilde{\varphi} \in L_{2,F}$$

for the functions $\varphi \in D = C_0^\infty(R^d)$ in (3.6) can be extended to the *unitary* correspondence

$$W \leftrightarrow L_{2,F} = \widetilde{W}, \tag{3.7}$$

since the Fourier transform $\tilde{D}$ is dense in J with it's closure $[\tilde{J}] = L_{2,F}$. Thus, for the space W we have it's Fourier transform $\widetilde{W} = L_{2,F}$.

The norm $\|\varphi\|_W$ is continuous over the Schwartz space $D = C_0^\infty(R^d)$, since the norm $\|\tilde{\varphi}\|_{L_{2,F}}$ is continuous with respect to convergence in the space $\tilde{J} = J$. Hence, we have W as the W-type space,

$$W = [D] \subseteq D^*,$$

with all $u \in W$ as the Schwartz distributions which appear by the inverse generalized Fourier transform of the corresponding $f = \tilde{u} \in L_{2,F}$. We are getting the following result.

THEOREM. *The space W with the Fourier transform $\widetilde{W} = L_{2,F}$ is of the W-type.* ∎

Let us turn on the dual space $X = W^*$. The Riesz formula

$$\langle \varphi, f \rangle_{L_{2,F}} = \int \varphi(\lambda)\overline{[F(\lambda)f(\lambda)]}\, d\lambda = (\varphi, \overline{Ff}), \qquad \varphi \in J,$$

describes all elements $\overline{Ff} \in L_{2,F}^*$ from the dual space $L_{2,F}^*$ as Schwartz distributions, represented by means of the corresponding functions

$$Ff = F(\lambda)f(\lambda), \qquad \lambda \in R^d,$$

with $f \in L_{2,F}$. Obviously, these are *all* functions $g = g(\lambda)$, $\lambda \in R^d$ such that

$$g \cdot F^{-1/2} \in L_2(R^d).$$

Assuming that we can apply $1/F$ as the weight function to form the corresponding space $L_{2,1/F}$, we see that this space is exactly formed with all functions $g = Ff$, $f \in L_{2,F}$. Moreover, according to the equality

$$\langle \varphi, f \rangle_{L_{2,F}} = \langle F\varphi, Ff \rangle_{L_{2,1/F}}$$

with $\varphi, f \in L_{2,F}$, we have the Hilbert space $L_{2,1/F}$ as the dual space to $L_{2,F}$,

$$L_{2,F}^* = L_{2,1/F}.$$

Taking in account the unitary correspondence (3.7), we obtain the similar one for the dual spaces

$$X \leftrightarrow L_{2,1/F} = \widetilde{X}. \tag{3.8}$$

In particular, the dual space $X = W^*$ is of the W-type,

$$X = [D] \subseteq D^*.$$

We formulate our results as follows.

THEOREM. *The dual space $X = W^*$ is of the W-type with the Fourier transform $\widetilde{X} = L_{2,1/F}$.* ■

With application of our scheme (3.6)–(3.8) we can, for example, describe the Sobolev spaces

$$W = \overset{\circ}{W}{}_2^p(R^d), \qquad X = \overset{\circ}{W}{}_2^{-p}(R^d)$$

with the integer $p \geq 0$ by the Fourier transform as

$$\widetilde{W} = L_{2,F}, \qquad \widetilde{X} = L_{2,1/F} \tag{3.9}$$

taking an appropriate weight-function[1]

$$F(\lambda) \asymp 1 + |\lambda|^{2p}, \qquad \lambda \in R^d. \tag{3.10}$$

We explain, that with $\lambda = (\lambda_1, \ldots, \lambda_d) \in R^d$, $\lambda^k = \lambda_1^{k_1} \ldots \lambda_d^{k_d}$, $k = (k_1, \ldots, k_d)$ the norm $\|u\|_p$ in $\overset{\circ}{W}{}_2^p$ can be given as

$$\|u\|_p^2 = \sum_{|k| \leq p} \|\partial^k u\|_{L_2}^2 = \sum_{|k| \leq p} \int |\tilde{u}(\lambda)|^2 \lambda^{2k} \, d\lambda$$

so $W = \overset{\circ}{W}{}_2^p(R^d)$ fits the scheme (3.6)–(3.8) with

$$F(\lambda) = \sum_{|k| \leq p} \lambda^{2k} \asymp 1 + |\lambda|^{2p}, \qquad \lambda \in R^d,$$

and by taking any other F of the type (3.10) we are getting the same space $L_{2,F} = \widetilde{W}$ which will be just with other norm, equivalent to the former one.

---

[1] The equivalence $\alpha \asymp \beta$ for variables α, β means $0 < c_1 \leq \frac{\alpha}{\beta} \leq c_2 < \infty$.

From the Fourier representation (3.9) we see, in particular, that $u \in \overset{\circ}{W}{}_2^p(R^d)$ *if and only if it is with generalized derivatives*

$$\partial^k u \in L_2(R^d), \qquad |k| \le p;$$

to explain, this is equivalent to the condition

$$\widetilde{\partial^k u} \in L_2(R^d), \qquad |k| \le p,$$

for the Fourier transform $\widetilde{\partial^k u} = (i\lambda)^k \tilde{u}(\lambda)$, $\lambda \in R^d$. And $x \in \overset{\circ}{W}{}_2^{-p}(R^d)$ *if and only if it is a Schwartz distribution with the Fourier transform such that*

$$\int |\tilde{x}(\lambda)|^2 (1 + |\lambda|^{2p})^{-1}\, d\lambda < \infty.$$

A very similar description holds true for the anisotropic Sobolev spaces. Namely, we have the Fourier representation (3.9) with the corresponding weight-function

$$F(\lambda) \asymp 1 + \sum_{j=1}^{m} |\lambda_j|^{2p_j}, \qquad \lambda \in R^d, \tag{3.10$'$}$$

where

$$\lambda_j \in R^{d_j}, \qquad \sum_{j=1}^{m} d_j = d, \qquad \lambda = (\lambda_1, \dots, \lambda_m) \in R^d$$

characterize anisotropic structure of these spaces with the multiindex $p = (p_1, \dots, p_m)$. We have $u \in \overset{\circ}{W}{}_2^p(R^d)$ *if and only if*

$$\partial^k u \in L_{2,F}, \qquad |k_j| \le p_j, \qquad j = 1, \dots, m,$$

for the corresponding group of variables $t_j \in R^{d_j}$ *in* $t = (t_1, \dots, t_m) \in R^d$. *And* $x \in \overset{\circ}{W}{}_2^{-p}(R^d)$ *if and only if it is a Schwartz distribution with the Fourier transform such that*

$$\int |\tilde{x}(\lambda)|^2 \left(1 + \sum_{j=1}^{m} |\lambda_j|^{2p_j}\right)^{-1} d\lambda < \infty. \quad \blacksquare$$

3.3. The equivalent Sobolev spaces

We set $W = \overset{\circ}{W}(T)$ to indicate a Hilbert space $W = [C_0^\infty(T)]$ in a region $T \subseteq R^d$. Having the Fourier representation (3.6) of it's scalar product with an appropriate weight function F, we treat this $\overset{\circ}{W}(T)$ as the subspace in the corresponding $\overset{\circ}{W}(R^d)$ of the W-type with the Fourier transform

$$\widetilde{\overset{\circ}{W}(R^d)} = L_{2,F}.$$

This approach let us, in particular, to treat elements $u \in \overset{\circ}{W}(T)$ as Schwartz distributions not only in the region $T \subseteq R^d$ but in the whole R^d; here we have in mind that

$$\overset{\circ}{W}(T) \subseteq \overset{\circ}{W}(R^d) = [D] \subseteq D^* \tag{3.11}$$

and the elements $u \in \overset{\circ}{W}(T)$ can be treated as the corresponding generalized functions

$$u = (\varphi, u), \qquad \varphi \in D = C_0^\infty(R^d).$$

We consider of how the spaces $\overset{\circ}{W}(T)$ relate to each other for different weight functions F, having the norm

$$\|u\|_W = \|\tilde{u}\|_{L_{2,F}}, \qquad u \in \overset{\circ}{W}(T).$$

Namely, the very question is on the equivalence of the different norms

$$\|\tilde{u}\|^2_{L_{2,F_1}} = \int |\tilde{u}(\lambda)|^2 F_1(\lambda)\, d\lambda \asymp \|\tilde{u}\|^2_{L_{2,F_2}} =$$

$$\int |\tilde{u}(\lambda)|^2 F_2(\lambda)\, d\lambda, \qquad u \in C_0^\infty(T) \tag{3.12}$$

giving us actually the same space $\overset{\circ}{W}(T) = [C_0^\infty(T)]$.

Considering the question on the *equivalence condition*

$$\|u\|_{W_1} \asymp \|u\|_{W_2}, \qquad u \in C_0^\infty(T)$$

of different norms in the W-type spaces $W = [C_0^\infty(T)]$, it is worthy to note at the very beging, that under the condition above, we have the equivalent norms

$$\|x\|_{X_1} = \sup_{\|\varphi\|_{W_1} \le 1} |(\varphi, x)| \asymp \sup_{\|\varphi\|_{W_2} \le 1} |(\varphi, x)| = \|x\|_{X_2}$$

in the dual spaces $X_1 = W_1^*$, $X_2 = W_2^*$. ■

Let us turn on our scheme (3.6)–(3.8).

Obviously, in a case of $T = R^d$ the equivalence (3.12) takes place if and only if we deal with the equivalent weight-functions

$$F_1 \asymp F_2,$$

that is $F_1(\lambda) \asymp F_2(\lambda)$ for almost all $\lambda \in R^d$; note, that in (3.12) we can go to any limit function $w = \lim |\tilde{u}|^2 \geq 0$, $\int w(\lambda)F(d\lambda)\, d\lambda < \infty$, considering $F = F_1, F_2$.

From now we consider a case of the *bounded* region $T \subseteq R^d$ assuming the weight functions to be as

$$F(\lambda) \geq c_0 > 0, \quad F(\lambda) = F_1(\lambda) \geq cF_2(\lambda), \qquad \lambda \in R^d. \tag{3.13}$$

THEOREM. *The norm equivalence* (3.12) *takes place for the weight functions* F_1, F_2 *with the difference* $F_1 - F_2$ *having generalized Fourier transform as locally square-integrable function.*

PROOF. Let us take the reverse Fourier transform $b(t)$, $t \in R^d$, of the difference $F_1(\lambda) - F_2(\lambda)$, $\lambda \in R^d$, that is

$$F_1 - F_2 = \tilde{b} = \bar{\tilde{b}},$$

and set

$$b(s,t) = b(s+t), \qquad (s,t) \in T \times T$$

extending this function from the bounded region $T \times T \subseteq R^d \times R^d$ on the whole space $R^d \times R^d$ as some square integrable function

$$b(s,t), \qquad (s,t) \in R^d \times R^d.$$

For any $u \in C_0^\infty(T)$, according to the product-convolution formulas and the Parseval equality, we get

$$\|\tilde{u}\|^2_{L_{2,F_1}} - \|\tilde{u}\|^2_{L_{2,F_2}} = \int [\tilde{u}(\lambda)\overline{\tilde{u}(\lambda)}]\overline{\tilde{b}(\lambda)}\, d\lambda =$$

$$= \frac{1}{(2\pi)^{d/2}} \int \left[\int u(s-t)\overline{u(-t)}\, dt \right] \overline{b(s)}\, ds =$$

$$= \frac{1}{(2\pi)^{d/2}} \iint [u(s)\overline{u(-t)}]\overline{b(s+t)}\, ds\, dt =$$

$$= \frac{1}{(2\pi)^{d/2}} \iint [u(s)\overline{u(-t)}]\overline{b(s,t)}\, ds\, dt =$$

$$= \frac{1}{(2\pi)^{d/2}} \iint\limits_{R^d \times R^d} [\tilde{u}(\lambda)\overline{\tilde{u}(\mu)}]\overline{\tilde{b}(\lambda,\mu)}\, d\lambda\, d\mu$$

with the Fourier transform $\tilde{u}(\lambda)\overline{\tilde{u}(\mu)}$, $(\lambda, \mu) \in R^d \times R^d$ of the product $u(s)\overline{u(-t)}$, $(s,t) \in R^d \times R^d$. Under the condition (3.13) with $F(\lambda) \geq c_0 > 0$ we have

$$a(\lambda, \mu) = \frac{\tilde{b}(\lambda, \mu)}{F(\lambda)F(\mu)}, \qquad (\lambda, \mu) \in R^d \times R^d,$$

in the known type space $L_{2,F\times F}$ with the weight function $F(\lambda)F(\mu)$, $(\lambda, \mu) \in R^d \times R^d$, since

$$\iint\limits_{R^d \times R^d} |a(\lambda,\mu)|^2 F(\lambda)F(\mu)\, d\lambda\, d\mu \leq C \iint\limits_{R^d \times R^d} |\tilde{b}(\lambda,\mu)|^2\, d\lambda\, d\mu < \infty.$$

Hence, considering $\tilde{u} \times \overline{\tilde{u}} = \tilde{u}(\lambda)\overline{\tilde{u}(\mu)}$, $(\lambda, \mu) \in R^d \times R^d$, in the space $L_{2,F\times F}$, we obtain

$$\|\tilde{u}\|^2_{L_{2,F_1}} - \|\tilde{u}\|^2_{L_{2,F_2}} = \langle \tilde{u} \times \tilde{u}, a\rangle_{L_{2,F\times F}}.$$

It shows, that

$$\|\tilde{u}_n\|^2_{L_{2,F_1}} - \|\tilde{u}_n\|^2_{L_{2,F_2}} \to 0 \tag{3.14}$$

for any orthonormal system of functions $\tilde{u} = \tilde{u}_n$, $n = 1, 2, \ldots$, in $L_{2,F}$, since the corresponding functions $\tilde{u} \times \overline{\tilde{u}} = \tilde{u}_n \times \overline{\tilde{u}}_n$, $n = 1, 2, \ldots$, form the orthonormal system in $L_{2,F\times F}$. To finish the proof, we apply the following preposition.

LEMMA. *The norm equivalence* (3.12) *takes place under the conditions* (3.13), (3.14).

To prove it, we are to show that

$$\|\tilde{u}\|_{L_{2,F_1}} \leq C\|\tilde{u}\|_{L_{2,F_2}}, \qquad u \in C_0^\infty(T),$$

since under the condition (3.13) we have

$$\|\tilde{u}\|_{L_{2,F_1}} \geq c\|\tilde{u}\|_{L_{2,F_2}}.$$

We apply the known Riesz type representation

$$\langle \tilde{u}, \tilde{v}\rangle_{L_{2,F_1}} - \langle \tilde{u}, \tilde{v}\rangle_{L_{2,F_2}} = \langle \tilde{u}, A\tilde{v}\rangle_{L_{2,F}}, \qquad u, v \in C_0^\infty(T),$$

for the bilinear continuous form in the left-hand side by means of the corresponding bounded operator A in the Hilbert space $L_{2,F}$ with the weight function $F = F_1$. According to the condition (3.14), this *symmetric* operator A is compact, and therefore

$$\sup_{\|\tilde{u}\|_{L_{2,F}} \leq 1} \langle \tilde{u}, A\tilde{u} \rangle_{L_{2,F}} = \langle f, Af \rangle_{L_{2,F}} = \alpha \leq 1$$

is reached at an eigenfunction f of A, $\|f\|_{L_{2,F}} = 1$. Here the condition $\alpha < 1$ exactly means the norm equivalence (3.12), and it does take place. Indeed, the contrary case $\alpha = 1$ means that

$$\|f\|^2_{L_{2,F_2}} = \int |f(\lambda)|^2 F_2(\lambda)\, d\lambda = 0,$$

and this (with $F_2 > 0$ on some open set, say) could be only for $f(\lambda) \equiv 0$, since

$$f(\lambda) = \frac{1}{(2\pi)^{d/2}} \int_T e^{-i\lambda t} u(t)\, dt$$

with $u \in L_2(T)$ represents an entire analytical function of the complex variable λ; note, that under the condition (3.13) with $F(\lambda) \geq c_0 > 0$ we have

$$\widetilde{\overset{\circ}{W}}(T) \subseteq L_{2,F} \subseteq L_2(R^d)$$

and, correspondingly,

$$\overset{\circ}{W}(T) = [C_0^\infty(T)] \subseteq L_2(R^d).$$

The proof is over. ■

For example, let us consider any weight function F of the type

$$0 < \varliminf_{\lambda \to \infty} (1 + |\lambda|^{2p})^{-1} F(\lambda) < \varlimsup_{\lambda \to \infty} (1 + |\lambda|^{2p})^{-1} F(\lambda) < \infty. \quad (3.15)$$

Obviously, we can take another weight function $F = F_1$ of the type (3.10) which coincides with the former $F = F_2$ for sufficiently large $|\lambda|$, and with $F = F_1$ we have the Sobolev space

$$\overset{\circ}{W}(T) = \overset{\circ}{W}{}_2^p(T).$$

Here the difference $F_1 - F_2$ is the function vanishing out of some compact and it satisfies the condition of our theorem on the norms equivalence — see (3.13). Thus *the Sobolev space* $W = \overset{\circ}{W}{}_2^p(T)$ *in the bounded region* T *can be equiped with the equivalent norm*

$$\|u\|_W = \|\tilde{u}\|_{L_{2,F}} \asymp \|u\|_p$$

given by means of any weight function F of the type (3.15).

The similar result holds true for the *anisotropic Sobolev space* $W = \overset{\circ}{W}{}_2^p(T)$ with the multi-index $p = (p_1, \dots, p_m)$ and weight functions F of the type

$$0 < \varliminf_{\lambda\to\infty} \left(1+\sum_{j=1}^{m} |\lambda_j|^{2p_j}\right)^{-1} F(\lambda) \le \varlimsup_{\lambda\to\infty} \left(1+\sum_{j=1}^{m} |\lambda_j|^{2p_j}\right)^{-1} F(\lambda) < \infty \tag{3.15$'$}$$

which correspond to (3.10)′. ■

Let us now turn on our scheme (3.1)–(3.4) with the general differential operator

$$\mathcal{P} = \mathcal{P}(\partial) = \sum_{|k|\le n} a_k \partial^k \ge 0$$

having *constant* (real) coefficients. We apply the corresponding polynomial

$$\mathcal{P} = \mathcal{P}(i\lambda) = \sum_{|k|\le n} a_k (i\lambda)^k, \qquad \lambda \in R^d.$$

The positiveness of the operator $\mathcal{P} \ge 0$,

$$(\varphi, \mathcal{P}\bar{\varphi}) = \int |\tilde{\varphi}|^2 \mathcal{P}(i\lambda)\, d\lambda \ge 0, \qquad \varphi \in C_0^\infty(R^d),$$

can be extended on all functions $\varphi \in J$ with the Fourier transform $\tilde{\varphi}$ representing, in particular, all functions $\tilde{\varphi} \in C_0^\infty(R^d)$, and it shows, that the positiveness $\mathcal{P}(\partial) \ge 0$ can be characterized in terms of the corresponding polynomial as

$$F(\lambda) = \mathcal{P}(i\lambda) \ge 0, \qquad \lambda \in R^d. \tag{3.16}$$

Clearly, we can represent the scalar product (3.1) in a form (3.6) with the defined above weight function F.

Suppose that

$$\varliminf_{\lambda\to\infty} \mathcal{P}(i\lambda) > 0. \tag{3.17}$$

Then we can take another weight function $F = F_1$ such that $F(\lambda) = \mathcal{P}(i\lambda)$ for sufficiently large $|\lambda|$ and $F(\lambda) \ge c > 0$ for all $\lambda \in R^d$. With this $F = F_1$ and the former $F = F_2$, defined by (3.16), we have the difference $F_1 - F_2$ vanishing out of some compact, so we can apply our theorem on the norms equivalence — see (3.13). By this in our scheme (3.1)–(3.4) we are getting the corresponding space $W = \overset{\circ}{W}(T)$ in any *bounded* region $T \subseteq R^d$ with the norm

$$\|u\|_W \asymp \|\tilde{u}\|_{L_{2,F}},$$

such that

$$\|u\|_W \geq c\|u\|_{L_2}, \qquad u \in \overset{\circ}{W}(T) \in L_2(T). \tag{3.17$'$}$$

This semiboundness condition is stronger than the general condition (3.2). Thus, we are getting in our scheme (3.1)–(3.4) the W-type spaces

$$W = [C_0^\infty(T)], \qquad X = W^* = [C_0^\infty(T)]. \quad \blacksquare$$

Note that in the positiveness condition (3.16) for the general differential operator $\mathcal{P} = \mathcal{P}(\partial) \geq 0$ the dominating role plays the corresponding homogeneous form

$$\sum_{|k|=n} a_k (i\lambda)^k = i^n \sum_{|k|=n} a_k \lambda^k, \qquad \lambda \in R^d,$$

considered for $\lambda \to \infty$; obviously, the condition (3.16) implies

$$i^n \sum_{|k|=n} a_k \lambda^k \geq 0, \qquad \lambda \in R^d,$$

with the even $n = 2p$. The condition

$$(-1)^p \sum_{|k|=2p} a_k \lambda^k > 0, \qquad \lambda \neq 0, \tag{3.18}$$

defines the *elliptic polynomial* $\mathcal{P}(i\lambda) = \sum_{|k|\leq 2p} a_k (i\lambda)^k$, $\lambda \in R^d$, and the corresponding elliptic differential operator $\mathcal{P}(\partial) = \sum_{|k|\leq 2p} a_k \partial^k$. (For example, $\mathcal{P}(\partial) = (-1)\Delta^p$ with the *Laplace operator*

$$\Delta = \sum_{j=1}^{d} \frac{\partial^2}{\partial t_j^2}$$

corresponds to the polynomial $\mathcal{P}(i\lambda) = |\lambda|^{2p}$, $\lambda \in R^d$). Under ellipticity condition (3.18) we have

$$c_1 |\lambda|^{2p} \leq (-1)^p \sum_{|k|\leq 2p} a_k \lambda^k \leq c_2 |\lambda|^{2p}, \qquad \lambda \in R^d$$

(where $c_1, c_2 > 0$ are min, max of the homogeneous form considered for $|\lambda| = 1$) and with application of the weight function $F(\lambda) = \mathcal{P}(i\lambda)$, $\lambda \in R^d$, of the type (3.15) we are getting in any bounded region $T \subseteq R^d$ the Sobolev space $W(T) = \overset{\circ}{W}{}_2^p(T)$ with the equivalent norm

$$\|u\|_W = \|\tilde{u}\|_{L_{2,F}} \asymp \|u\|_p.$$

Consequently, in our scheme (3.1)–(3.4), associated with the differential operator $\mathcal{P} = \mathcal{P}(\partial)$, we have the Sobolev spaces

$$W = \overset{\circ}{W}{}_2^p(T), \qquad X = W^* = \overset{\circ}{W}{}_2^{-p}(T).$$

The very similar result holds true for the differential operator $\mathcal{P} = \mathcal{P}(\partial) \geq 0$, characterized by the condition (3.15)′ for the corresponding $F(\lambda) = \mathcal{P}(i\lambda)$, $\lambda \in R^d$; namely, in our scheme (3.1)–(3.4) we have the anisotropic Sobolev spaces

$$W = \overset{\circ}{W}{}_2^{(p_1,\dots,p_m)}, \qquad X = W^* = \overset{\circ}{W}{}_2^{-(p_1,\dots,p_m)}(T).$$

For example, taking the differential operator

$$\mathcal{P}(\partial) = -\frac{\partial^2}{\partial r^2} + Q(\partial) \tag{3.19}$$

where $Q = \sum_{|k| \leq 2p} a_k \partial^k$ is the elliptic operator over the variable $t_1 \in R^{d-1}$ and $\partial^2/\partial r^2$ is over the remaining variable $r = t_2 \in R^1$, we have in our scheme (3.1)–(3.4) the anisotropic Sobolev spaces

$$W = \overset{\circ}{W}{}_2^{(p,1)}(T), \qquad X = W^* = \overset{\circ}{W}{}_2^{-(p,1)}(T)$$

with the multi-index $(p, 1)$; we explain, that taking φ_r for $\varphi \in C_0^\infty(T)$ with the variable $r \in R^1$ fixed, we have

$$(\varphi, \mathcal{P}\varphi) = \left(\varphi, \frac{\partial^2}{\partial r^2}\varphi\right) + \int_{R_1} (\varphi_r, Q\varphi_r)\, dr \asymp$$

$$\asymp \left\| \frac{\partial}{\partial r}\varphi \right\|_{L_2(R^d)}^2 + \int_{R_1} \sum_{|k| \leq p} \|\partial^k \varphi_r\|_{L_2(R^{d-1})}^2\, dr =$$

$$= \left\| \frac{\partial}{\partial r}\varphi \right\|_{L_2(R^d)}^2 + \sum_{|k| \leq p} \|\partial^k \varphi\|_{L_2(R^d)}^2. \quad \blacksquare$$

3.4. Multiplicators and localization of generalized functions

We are to characterize multiplicators in the W-type spaces $W = [D]$, $X = W^* \subseteq D^*$.

For example, multiplication of any generalized function $x \in D^*$ by infinitely differentiable function w is defined as

$$wx = (\varphi, wx) = (w\varphi, x), \qquad \varphi \in D = C_0^\infty(T),$$

and then the question is whether for $x \in X \subseteq D^*$ this multiplication results in the function $wx \in D^*$ belonging to X.

Let us assume that the function w is such that for the product defined by the above formula $wx \in X$ for all $x \in X$ (in particular, it implies $w\varphi \in W$ for all $\varphi \in D$). Let us show that the linear operator of multiplication $x \to wx$ is closed in X. From convergence $x_n \to x$, $wx_n \to y$ in X, we have weak convergence

$$\begin{array}{ccccc} (\varphi, wx_n) & = & (w\varphi, x_n) & & \\ \downarrow & & \downarrow & & \\ (\varphi, y) & = & (w\varphi, x) & = & (\varphi, wx), \qquad \varphi \in D, \end{array}$$

and therefore $y = wx$. It is known that any closed linear operator defined on the Hilbert space is bounded; hence,

$$\|wx\|_X \le C\|x\|_X, \qquad x \in X.$$

This implies

$$|(w\varphi, x)| = |(\varphi, wx)| \le \|\varphi\|_W \|wx\|_X \le C\|\varphi\|_W \|x\|_X$$

and

$$\|w\varphi\|_W = \sup_{\|x\|_X \le 1} |(w\varphi, x)| \le C\|\varphi\|_W, \qquad \varphi \in D,$$

i.e., the operator of multiplication $\varphi \to w\varphi$, defined directly on D, is bounded in W. In turn, by virtue of the condition of boundness of this operator on D in $W = [D]$,

$$\|w\varphi\|_W \le C\|\varphi\|_W, \qquad \varphi \in D, \tag{3.20}$$

for any $x \in X$ and $wx = (w\varphi, x)$, $\varphi \in D$, we obtain

$$|(w\varphi, x)| \le \|w\varphi\|_W \|x\|_X \le C\|\varphi\|_W \|x\|_X.$$

This implies that $wx \in X = W^*$ with the norm

$$\|wx\|_X \leq C\|x\|_X. \tag{3.20$'$}$$

Note that the multiplication operator $w: \varphi \to w\varphi$ on $\varphi \in D$, being bounded in W, can be uniquely extended on the whole space $W = [D]$ as

$$wu = \lim w\varphi \tag{3.21}$$

for all $u = \lim \varphi$ in W. And according to

$$(x, wu) = (wu, x) = \lim(w\varphi, x) =$$

$$= \lim(\varphi, wx) = (u, wx) = (wx, u), \qquad x \in D,$$

in the case $X = [D]$ and $W \subseteq D^*$, the multiplication operator defined in (3.21) acts on limiting elements $u = \lim \varphi$ as on generalized functions $u \in D^*$.

Let us summarize the above reasoning.

THEOREM. *The multiplication operator is well-defined on the spaces $W = [D]$, $X = W^*$ if and only if it is bounded on D in W.* ■

For example, let us turn on the Sobolev spaces $W = \overset{\circ}{W}{}_2^p(T)$, $X = \overset{\circ}{W}{}_2^{-p}(T)$. Obviously, the condition (3.20) is satisfied for any function $w \in C_b^p(T)$, bounded together with its derivatives $\partial^k w$, $|k| \leq p$. To see this, we recall the Leibnith formula

$$\partial^k(w\varphi) = \sum_{l \leq k} \frac{k!}{l!\,(l-k)!} \partial^l w \partial^{k-l} \varphi$$

(with $l_i \leq k_i$ for components of multiindices $l \leq k$) by virtue of which

$$\|w\varphi\|_W^2 \asymp \sum_{|k| \leq p} \|\partial^k(w\varphi)\|_{L_2}^2 \leq C \sum_{|k| \leq p} \|\partial^k \varphi\|_{L_2}^2 \leq C\|\varphi\|_W^2, \qquad \varphi \in D.$$

The similar type of multiplicators w with bounded derivatives $\partial^{k_j} w$, $|k_j| \leq p_j$, $j = 1, \dots, m$, we have in the anisotropic Sobolev spaces

$$W = \overset{\circ}{W}{}_2^{(p_1,\dots,p_m)}(T), \qquad X = \overset{\circ}{W}{}_2^{-(p_1,\dots,p_m)}(T);$$

in particular, we can apply multiplication by $w \in C_0^\infty(T)$. A wider class of W-type spaces can be characterized by the condition

$$c \sum_k \int_{T_{\text{loc}}} |\partial^k \varphi|^2 \, dt \leq \|\varphi\|_W^2 \leq C \sum_k \|\partial^k \varphi\|_{L_2(T)}^2, \qquad \varphi \in D, \tag{3.22}$$

in any bounded region $T_{\text{loc}} \subseteq T$ where together with the derivative $\partial^k\varphi$ all lower derivatives $\partial^l\varphi$, $l \leq k$ ($l_j \leq k_j$ for multiindexes) are presented. By virtue of condition (3.22), in $W = [D]$, $X = W^*$ multiplication by any function $w \in C_0^\infty(T)$ is well-defined. Indeed, taking $T_{\text{loc}} \supseteq \operatorname{supp} w$, we obtain the multiplication operator boundness,

$$\|w\varphi\|_W^2 \leq C\sum_k \|\partial^k(w\varphi)\|_{L_2}^2 \leq$$

$$\leq C\sum_k \int_{T_{\text{loc}}} |\partial^k\varphi|^2\,dt \leq C\|\varphi\|_W^2, \qquad \varphi \in D. \quad \blacksquare$$

Let us consider the spaces $W = [D]$, $X = W^*$ in our scheme (3.6)–(3.8) with the corresponding weight function F. Suppose, the function $F(\lambda)$, $\lambda \in R^d$, is such that

$$F(\lambda + \mu) \leq G(\mu)F(\lambda) \tag{3.23}$$

for all $\mu \in R^d$; this property holds true, for example, for weight functions F of the type (3.10), (3.10)′ leading to the Sobolev spaces.

The product $w\varphi$ has the convolution $(2\pi)^{d/2}\tilde{w} * \tilde{\varphi} = \widetilde{w\varphi}$ as its Fourier transform and the boundness condition (3.20) can be given in the form

$$\|\tilde{w} * \tilde{\varphi}\|_{L_{2,F}} \leq C\|\tilde{\varphi}\|_{L_{2,F}}, \qquad \varphi \in D.$$

Suppose, the Fourier transform $\tilde{w}$ is such that the function $f(\mu) = |\tilde{w}|G(\mu)^{1/2}$ is integrable,

$$\int |\tilde{w}|G(\mu)^{1/2}\,\partial\mu < \infty. \tag{3.24}$$

Then, using the inequality $F(\lambda) \leq F(\lambda - \mu)G(\mu)$ and assuming $\int f(\lambda)\,d\lambda = C$, we obtain

$$\|\tilde{w} * \tilde{\varphi}\|_{L_{2,F}}^2 = \int \left| \int \tilde{\varphi}(\lambda - \mu)\tilde{w}(\mu)F(\lambda)^{1/2}\,d\mu \right|^2 d\lambda \leq$$

$$\leq \int \left| \int \tilde{\varphi}(\lambda - \mu)F(\lambda - \mu)^{1/2} f(\mu)\,d\mu \right|^2 d\lambda \leq$$

$$\leq C \int\int |\tilde{\varphi}(\lambda - \mu)|^2 F(\lambda - \mu) f(\mu)\,d\mu\,d\lambda =$$

$$= C\int \left[\int |\tilde{\varphi}(\lambda - \mu)|^2 F(\lambda - \mu)\,d\lambda \right] f(\mu)\,d\mu = C^2\|\tilde{\varphi}\|_{L_{2,F}}^2.$$

Thus, the boundness condition (3.20) holds. Therefore, we obtained the following result.

THEOREM. *Multiplication by any function w satisfying the condition* (3.24) *is well-defined in the spaces $W = [D]$, $X = W^*$.*

Of course, this result is valid for all other W, X in a case of equivalent norms

$$\|\tilde{\varphi}\|_W^2 \asymp \|\tilde{\varphi}\|_{L_{2,F}}^2, \qquad \varphi \in D. \quad \blacksquare$$

Let us consider a general space $W = [D] \subseteq D^*$ such that the *unity approximators* $w_n = w(t/n)$, $n \to \infty$ (with $w \in C_0^\infty(R^d)$, $w = 1$ in some neighborhood of $t = 0$) as multiplication operators are bounded in W,

$$\|w_n \varphi\|_W \leq C\|\varphi\|_W, \qquad \varphi \in D = C_0^\infty(T).$$

Then localization of the functions $u \in W$ in subregion $S \subseteq T$ leads to the following result.

LEMMA. *The functions $u \in W$ with supports* $\operatorname{supp} u \subseteq S$ *belong to the subspace*

$$\overset{\circ}{W}(S) = [C_0^\infty(S)].$$

PROOF. In a case of compact support $\operatorname{supp} u$, we can take a multiplicator $w \in C_0^\infty(S)$, $w = 1$ in neighborhood of the compact $\operatorname{supp} u$, for which, in representation $u = \lim \varphi$ in W as a limit of functions $\varphi \in D = C_0^\infty(T)$, we have $u = wu = \lim w\varphi$. Thus $u = \lim w\varphi \in \overset{\circ}{W}(S)$ for $w\varphi \in C_0^\infty(S)$. In a case of any support $\operatorname{supp} u \subseteq S$, we can employ the unity approximators $w_n = w(t/n)$ for which $u = \lim w_n u \in \overset{\circ}{W}(S)$ as a limit of functions $w_n u \in \overset{\circ}{W}(S)$ with compact supports $\operatorname{supp} w_n u \subseteq \operatorname{supp} u \subseteq S$. Note that for the sequence of the multiplication operators w_n with bounded norm and obvious convergence $w_n\varphi \to \varphi$ in $D = C_0^\infty(T) \subseteq W$ we have

$$\|u - w_n u\|_W \leq \|u - \varphi\|_W + \|\varphi - w_n\varphi\|_W + \|w_n(\varphi - u)\|_W \to 0$$

with $\varphi \to u$ in W. $\blacksquare$

Let

$$W = [D], \qquad X = [D] = W^* \subseteq D^*$$

be W-type spaces in the region $T \subseteq R^d$ with multiplication on $w \in C_0^\infty(T)$ well-defined. We introduce a localized space $X[S]$, formed by all $x \in X$ as the Schwartz distributions

$$x = (\varphi, x), \qquad \varphi \in D = C_0^\infty(T),$$

with supports $\operatorname{supp} x \subseteq [S]$ in a closure of the region S, assuming that $[S] \subseteq T$. Thus, $X[S]$ is the certain Hilbert space of Schwartz distributions with the corresponding norm

$$\|x\|_X = \sup_{\|\varphi\|_W \leq 1} |(\varphi, x)|$$

where sup is taken over $\varphi \in C_0^\infty(T)$. By this definition, $X[S]$ seems to be dependent on the very region $T \supseteq [S]$, but it is not so for *bounded* S, at least.

Namely, let us take any bounded region T_{loc}, $[T_{\text{loc}}] \subseteq T$, such that $[S] \subseteq T_{\text{loc}}$, and apply $W_{\text{loc}} = [C_0^\infty(T_{\text{loc}})]$ as subspace in the former $W = [C_0^\infty(T)]$, considering $X_{\text{loc}} = W_{\text{loc}}^*$ as obtained by factorization of the former $X = W^*$ with respect to the corresponding norm

$$\|x\|_{X_{\text{loc}}} = \sup_{\|\varphi\|_{W_{\text{loc}}} \leq 1} |(\varphi, x)|, \qquad x \in X,$$

where sup is taken over $\varphi \in C_0^\infty(T_{\text{loc}})$. For a multiplicator $w \in C_0^\infty(T_{\text{loc}})$ with $w(t) = 1$, $t \in S$, and all $x \in X$ with supports $\operatorname{supp} x \subseteq [S]$ we have

$$\|x\|_{X_{\text{loc}}} \leq \|x\|_X = \sup_{\|\varphi\|_W \leq 1} |(w\varphi, x)| \leq$$

$$\leq C \sup_{\|w\varphi\|_W \leq 1} |(w\varphi, x)| \leq C\|x\|_{X_{\text{loc}}}.$$

Thus, for the localized space $X_{\text{loc}}[S]$ (associated with the W-type spaces $W_{\text{loc}} = [C_0^\infty(S)]$, $X_{\text{loc}} = W_{\text{loc}}^*$ in the region T_{loc}) we are getting the following result.

THEOREM. *The localized spaces $X_{\text{loc}}[S]$, $X[S]$ are the same having the equivalent norms*

$$\|x\|_{X_{\text{loc}}} \asymp \|x\|_X, \qquad x \in X_{\text{loc}}[S] = X[S]. \quad \blacksquare \tag{3.25}$$

In what follows later on, we actually apply a localized space of the type

$$X(S) = [C_0^\infty(S)] \tag{3.26}$$

which appears in the subregion $S \subseteq T$ as the closure of $C_0^\infty(S)$ in the corresponding W-type space $X = [D] = W^* \subseteq D^*$ with $D = C_0^\infty(T)$. (Obviously, $X(S)$ can be treated as the closure of $C_0^\infty(S)$ in the localized space $X[S]$.) ■

I.4. The Sobolev spaces

4.1. Some point-wise properties of generalized derivatives

At first, we are going to give a certain point-wise characterization of functions $u \in \overset{\circ}{W}{}_2^p(T)$ for the Sobolev spaces with the integer index $p \geq 0$. Considering $\overset{\circ}{W}{}_2^p(T) = [C_0^\infty(T)]$ in a region $T \subseteq R^d$ as a subspace in $\overset{\circ}{W}{}_2^p(R^d) = [C_0^\infty(R^d)]$ we turn on the Sobolev space $\overset{\circ}{W}{}_2^p(R^d)$.

Recall, the functions $u \in \overset{\circ}{W}{}_2^p(R^d)$ are characterized by a property of generalized derivatives $\partial^k u$, $|k| \leq p$, to be in $L_2(R^d)$-space.

For $|k| \leq p-1$, the function $f = \partial^k u$ and it's generalized derivative $g = \partial f/\partial r$ with respect to any coordinate $r = t_i$ in $t \in R^d$ are related to each other with the equality

$$-\int \frac{\partial \varphi}{\partial r} f\, dt = \int \varphi g\, dt, \qquad \varphi \in C_0^\infty(R^d).$$

Considering $t = (s, r)$ with $s \in R^{d-1}$, $r \in R^1$ we can treat $f, g \in L_2(R^d)$ in a way that $f(s,r)$, $g(s,r)$ for almost every $s \in R^{d-1}$ represent functions $f, g \in L_2(R^1)$ of the variable $r \in R^1$. And taking $\varphi(t) = x(s)w(r)$ with all $x \in C_0^\infty(R^{d-1})$, we see that for almost every $s \in R^{d-1}$

$$\int_a^b w'(r)\left[f - \int g\right] dr = 0, \qquad \int g = \int_a^r g(s,r)\, dr,$$

for a countable system of $w \in C_0^\infty(a,b)$ on a finite interval (a,b), say. Taking a *complete* system $\{w'\}$ of derivatives orthogonal to constant-functions in the $L_2(a,b)$-space, we obtain that for almost every $s \in R^{d-1}$

$$f(s,r) = \int_a^r g(s,r)\, dr + C, \qquad a \leq r \leq b.$$

We see, in particular, that here $f(s,r)$, $a \leq r \leq b$, is absolutely continuous with the corresponding derivative $\partial f/\partial r = g(s,r)$ well-defined for almost all r. With other hand, if $f = \partial^k u \in L_2(R^d)$

for almost all $s \in R^{d-1}$ represents absolutely continuous function of $r \in R^1$ with the corresponding derivative

$$\frac{\partial f(s,r)}{\partial r} = g(s,r) \in L_2(R^d), \tag{4.1}$$

then this g represents generalized derivative of f as well, which can be directly verified with integration by parts.

Thus, $u \in \overset{\circ}{W}{}_2^p(R^d)$ *if and only if all consequently functions* $f = \partial^k u$, $0 \leq |k| \leq p-1$, *are of the type* (4.1). ■

Let us turn on the space $C_0^p(R^d)$ of all functions $u = u(t)$, $t \in R^d$, with continuous derivatives $\partial^k u$, $|k| \leq p$, having compact supports $\operatorname{supp} u \subseteq R^d$. Obviously, $C_0^p(R^d) \subseteq \overset{\circ}{W}{}_2^p(R^d)$ and

$$[C_0^p(T)] = \overset{\circ}{W}{}_2^p(T).$$

We apply in R^d an appropriate transformation

$$t = \tau(\hat{t}) \tag{4.2}$$

with coordinate-wise functions $t_i = \tau_i(\hat{t})$, $i = 1, \ldots, d$, having bounded derivatives $\partial^k \tau_i$, $|k| \leq p$, and nondegenerate Jacobian $|\partial\tau| \geq c > 0$ such that the inverse transformation is of the same type.

Obviously, this type transformation gives one-to-one equivalently isometric mapping

$$\tau\colon\ C_0^p(R^d) \ni u \leftrightarrow \tau u \in C_0^p(R^d)$$

with the equivalent norms

$$\|u\|_p \asymp \|\tau u\|_p,$$

and it can be extended as one-to-one mapping of the corresponding Sobolev spaces,

$$\overset{\circ}{W}{}_2^p(R^d) = [C_0^p(R^d)] \ni u \leftrightarrow \tau u \in [C_0^p(R^d)] = \overset{\circ}{W}{}_2^p(R^d).$$

With this τ in $W = \overset{\circ}{W}{}_2^p(R^d)$ we can define the corresponding dual mapping

$$\tau^*\colon\ \overset{\circ}{W}{}_2^{-p}(R^d) \ni x \leftrightarrow \tau^* x \in \overset{\circ}{W}{}_2^{-p}(R^d)$$

in the dual space $X = W^* = \overset{\circ}{W}{}_2^{-p}(R^d)$ as

$$\tau^* x = (\tau^* x, u) = (x, \tau u), \qquad u \in W.$$

Obviously, we have

$$\|\tau^* x\|_{-p} = \sup_{\|u\|_p \le 1} (\tau^* x, u) \asymp$$

$$\asymp \sup_{\|\tau u\|_p \le 1} (x, \tau u) = \|x\|_{-p}, \qquad x \in X. \quad \blacksquare$$

We are going to consider certain properties of functions $u \in \overset{\circ}{W}{}_2^p(R^d)$ related to a boundary $\Gamma = \partial T$ of complementary to each other regions $T = T_-$ and $T_+ = R^d \setminus [T]$. Let us assume that the boundary Γ is *regular* as follows. It can be covered by local neighborhoods $\{O_\alpha\}$ in R^d,

$$\Gamma \subseteq \cup O_\alpha = \Gamma^\varepsilon,$$

and locally, in every neighborhood $O = O_\alpha$, described as a "hyperplane"

$$\Gamma = \{t = \tau(s, r)\colon\ s \in R^{d-1},\ r = 0\} \tag{4.3}$$

by appropriate coordinates

$$t = \tau(\hat{t}), \qquad \hat{t} = (s, r) \in R^{d-1} \times R^1$$

of the type (4.2) which fit a nontangent vector field $l = l(t)$, $t \in \Gamma$, in a way that

$$u^{(k)}(t) = \frac{\partial^k}{\partial r^k} u(\tau(s,r)) = \frac{\partial^k}{\partial l^k} u(t), \qquad t \in \Gamma^\varepsilon, \tag{4.3$'$}$$

represent indicated derivatives along the given $l = l(t)$, $t \in \Gamma$, in the boundary neighborhood $\Gamma^\varepsilon = \cup O_\alpha$.

The given in (4.1) pointwise characterization of the functions $u \in \overset{\circ}{W}{}_2^p(R^d)$ shows that there are absolutely continuous derivatives

$$\frac{\partial^k}{\partial r^k} u(\tau(s,r)), \qquad r \in R^1 \quad (k = 0, \dots, p-1)$$

for almost every $s \in R^{d-1}$, so there is their *boundary trace*

$$u^{(k)}(t) = \frac{\partial^k}{\partial r^k} u(\tau(s,0)) = \frac{\partial^k}{\partial l^k} u(t), \qquad t \in \Gamma \quad (k = 0, \dots, p-1) \tag{4.4}$$

on the boundary at almost every $t \in \Gamma$ with respect to the corresponding Lebesgue measure on Γ.

LEMMA. *Let* $u \in \overset{\circ}{W}{}_2^p(R^d)$ *has the zero boundary trace*

$$u^{(k)}(t) = 0, \qquad t = (s, 0) \in \Gamma \quad (k = 0, \dots, p-1) \tag{4.5}$$

for almost all $t \in \Gamma$ *with respect to the Lebesgue measure on* Γ*; then* $u \in \overset{\circ}{W}{}_2^p(R^d \setminus \Gamma)$.

We are going to prove it alltogether with another result.

THEOREM (on localization). *The subspace* $\overset{\circ}{W}{}_2^p(T) \subseteq \overset{\circ}{W}{}_2^p(R^d)$ *consists of all functions* $u \in \overset{\circ}{W}{}_2^p(R^d)$ *with supports* $\operatorname{supp} u \subseteq [T]$ *in the closure of* T.

PROOF. Obviously, $\operatorname{supp} u \subseteq [T]$ for all $u \in \overset{\circ}{W}{}_2^p(T)$, since $\overset{\circ}{W}{}_2^p(T) = [C_0^\infty(T)]$ is the closure of $C_0^\infty(T)$ and for any $x \in C_0^\infty(T_+)$ in the complementary region $T_+ = R^d \setminus [T]$ we have

$$(x, u) = \lim_{\varphi \to u} (x, \varphi) = 0$$

with $\varphi \in C_0^\infty(T)$. To prove, that the condition $\operatorname{supp} u \subseteq [T]$ implies $u \in \overset{\circ}{W}{}_2^p(T)$, we consider at first a case of the half-space

$$T = T_- = R_-^d = \{t = (s, r) \colon \ s \in R^{d-1}, \ r > 0\}.$$

Obviously, any $u \in \overset{\circ}{W}{}_2^p(R^d)$, which vanishes for $r > 0$, can be shifted to $u_h(s, r) = u(s, r + h) \in \overset{\circ}{W}{}_2^p(R^d)$ with $\operatorname{supp} u_h \subseteq R_-^d$. According to our lemma on localization (see p. 42), we have $u_h \in \overset{\circ}{W}{}_2^p(T)$. For all generalized derivatives $\partial^k u_h = (\partial^k u)_h$, $|k| \le p$, we have

$$\|\partial^k u_h - \partial^k u\|_{L_2(R^d)} \to 0, \qquad h \to 0,$$

and we see that $u = \lim_{h \to 0} u_h$ is a limit of the functions $u_h \in \overset{\circ}{W}{}_2^p(T)$, so $u \in \overset{\circ}{W}{}_2^p(T)$. Now we consider a general case. Any $u \in \overset{\circ}{W}{}_2^p(R^d)$ can be approximated by wu with the unity approximators $w \in C_0^\infty(R^d)$ on R^d. This let us consider $u \in \overset{\circ}{W}{}_2^p(R^d)$ with compact support $\operatorname{supp} u \subseteq [T]$. We have

$$u = \left[u - \sum_\alpha w_\alpha u\right] + \sum_\alpha w_\alpha u,$$

taking a proper *unity representation*

$$1 = \sum w_\alpha(t), \qquad t \in \Gamma \cap \operatorname{supp} u,$$

on Γ with a finite number of $w_\alpha \in C_0^\infty(O_\alpha)$, localized in the corresponding neighborhoods O_α of the local coordinates; recall, in every neighborhood O_α the boundary Γ is represented by the corresponding hyperplane — see (4.3), (4.3)'. As we know, a difference $u - \sum_\alpha w_\alpha u$ represents a function from $\overset{\circ}{W}{}_2^p(T)$, having it's support in the very region T. And with application of the local coordinates, as was actually shown above, we have $w_\alpha u \in \overset{\circ}{W}{}_2^p(T)$. Therefore, $u \in \overset{\circ}{W}{}_2^p(T)$ as well, and the theorem proof is over.

Let us return to the lemma. As in the theorem proof, at first we consider a case of the half-space $T = T_- = R_-^d$. Taking in account the condition (4.5), we see that, alltogether with the very function $u \in \overset{\circ}{W}{}_2^p(R^d)$, the functions

$$u_-(t) = \begin{cases} u(t), & t \in T_-, \\ 0, & t \in T_+, \end{cases} \qquad u_+(t) = \begin{cases} 0, & t \in T_-, \\ u(t), & t \in T_+, \end{cases} \tag{4.6}$$

fit our pointwise characterization of the Sobolev space $\overset{\circ}{W}{}_2^p(R^d)$ — see (4.1), that is

$$u_-, u_+ \in \overset{\circ}{W}{}_2^p(R^d).$$

We have $\operatorname{supp} u_- \subseteq [T_-]$, $\operatorname{supp} u_+ \subseteq [T_+]$ and, according to the theorem,

$$u_- \in \overset{\circ}{W}{}_2^p(T_-), \qquad u_+ \in \overset{\circ}{W}{}_2^p(T_+).$$

Hence,

$$u = u_- + u_+ \in \overset{\circ}{W}{}_2^p(T_- \cup T_+).$$

Thus, we have proved our lemma for a case of $T = R_-^d$. And for a general case we can go further in the very same way as in the theorem proof. ■

We are to note that shown above properties of the localized functions $u \in \overset{\circ}{W}{}_2^p(R^d)$ can be easily verified for the boundary Γ with singularities of the rectangular type as in the region

$$T = R_-^d(t_1, \dots, t_m) = \{t\colon\ t_j < 0,\ j = 1, \dots, m\}$$

of all $t = (t_1, \dots, t_d) \in R^d$ with the indicated coordinates $t_j < 0$, $j = 1, \dots, m$. We have in mind that $u \in \overset{\circ}{W}{}_2^p(R^d \setminus \Gamma)$ *if it has the zero boundary trace*, and $u \in \overset{\circ}{W}{}_2^p(T)$ *if* $\operatorname{supp} u \subseteq [T]$.

And these properties hold true for the general region $T \subseteq R^d$ which boundary $\Gamma = \partial T$ is *piecewise regular*, to be more precise, Γ can be covered with local neighborhoods $O_\alpha \subseteq R^d$ where either the boundary Γ locally is described as "hyperplane" (4.3) or the very region T locally, by means of appropriate local coordinates $t = \tau(\hat{t})$ of the type (4.2), is described in the boundary neighborhood as the "rectangular"

$$T = \{t = \tau(\hat{t}) \colon \hat{t} \in R_-^{d-1}(\hat{t}_1, \dots, \hat{t}_m)\} \tag{4.7}$$

For a proof one can go along the same way as in our theorem on localization.

4.2. Test functions spaces

We are going to treat a closure $X = [D]$ of the Schwartz test functions space $D = C_0^\infty(R^d)$ with respect to the norm

$$\|x\|_{-p}^2 = \int_{R^d} |\tilde{x}(\lambda)|^2 (1 + |\lambda|^{2p})^{-1}\, d\lambda$$

as the certain *test functions space.* Recall, we have here the Sobolev space $X = \overset{\circ}{W}{}_2^{-p}(R^d)$ of Schwartz distributions

$$x = (\varphi, x) \equiv (x, \varphi), \qquad \varphi \in D = C_0^\infty(R^d),$$

which can be described by it's Fourier transform $\widetilde{X} = L_{2,1/F}$ as the known $L_{2,1/F}$-space with the weight-function $1/F = (1 + |\lambda|^{2p})^{-1}$, $\lambda \in R^d$ — see (3.9), (3.10).

By means of $x \in X$ we can test any generalized random function $\xi = (\varphi, u)$, $\varphi \in D$, meansquare continuous with respect to the norm $\|\varphi\|_{-p}$ — the corresponding result can be obtained as the limit

$$(x, \xi) = \lim_{\varphi \to x} (\varphi, \xi)$$

— cf. (1.3); in particular, we can test all $u \in W = \overset{\circ}{W}{}_2^p(R^d)$ by means of $x = (x, u)$, $u \in W$, as elements $x \in X = W^*$ of the dual space.

Considering the region $T \subseteq R^d$, it seems natural to introduce the *test functions space*

$$X(T) = [C_0^\infty(T)]$$

as a closure of all Schwartz test functions $x \in C_0^\infty(T)$ with respect to the norm $\|x\|_{-p}$.

THEOREM. *The space* $X(T) = [C_0^\infty(T)]$ *consists of all Schwartz distributions* $x \in X$ *with supports* $\operatorname{supp} x \subseteq [T]$ *in the closure of* T.

PROOF. Applying our general theorem on localization to the region $T^+ = R^d \setminus [T]$ with the closure $[T^+] = T^c$ as a complement of the region T, we have

$$u \in \overset{\circ}{W}{}_2^p(T^+) = [C_0^\infty(T^+)]$$

for any $u \in W = \overset{\circ}{W}{}_2^p(R^d)$, $\operatorname{supp} u \subseteq T^c$, i.e., for any u which annulates all $x \in C_0^\infty(T)$ in the dual space $X = W^*$. Here,

$$u = \lim \varphi, \qquad \varphi \in C_0^\infty(T^+),$$

is such that

$$(x, u) = \lim(x, \varphi) = 0, \qquad \operatorname{supp} x \subseteq [T],$$

for all $x \in X$ with supports in the closure $[T]$. Regarding $u \in W$ as elements $u = (x, u)$, $x \in X$ of the dual space $W = X^*$, we see that any $u \in W$ which annulates all $x \in C_0^\infty(T)$, annulates the total space $X[T] \subseteq X$ of all $x \in X$, $\operatorname{supp} x \subseteq [T]$ as well; it shows that $X[T] = [C_0^\infty(T)] = X(T)$ and proves the theorem. ■

Let us turn on our scheme in (3.1) with any differential operator $\mathcal{P} \geq 0$ which gives the equivalent norm

$$\|\varphi\|_W = (\varphi, \mathcal{P}\bar{\varphi})^{1/2} \asymp \|\varphi\|_p, \qquad \varphi \in C_0^\infty(R^d),$$

in the Sobolev space $W = [C_0^\infty(R^d)] = \overset{\circ}{W}{}_2^p(R^d)$. Recall, we have it's dual space $X = \overset{\circ}{W}{}_2^{-p}(R^d)$ as

$$X = \mathcal{P}W$$

with the corresponding norm

$$\|\mathcal{P}u\|_X = (u, \mathcal{P}\bar{u})^{1/2} \asymp \|\mathcal{P}u\|_{-p}, \qquad u \in W.$$

Dealing with the test functions space $X(T) = [C_0^\infty(T)]$ in the region $T \subseteq R^d$, which consists of all $x \in X$ with $\operatorname{supp} x \subseteq [T]$, we have $\mathcal{P}C_0^\infty(T) \subseteq X(T)$. What is a complement to $\mathcal{P}C_0^\infty(T)$ in $X(T)$? It seems remarkable that we can take a *direct complement*

$$X(\Gamma) = \{x \in X(T) \colon \operatorname{supp} x \subseteq \Gamma\}$$

formed with all *boundary test functions* $x \in X(T)$ having supports $\operatorname{supp} x \subseteq \Gamma$ on the boundary $\Gamma = \partial T$.

THEOREM. *The direct sum representation*

$$X(T) = [\mathcal{P}C_0^\infty(T)] \oplus X(\Gamma) \tag{4.8}$$

holds true.

PROOF. We recall the dual Riesz formula

$$\langle \varphi, u \rangle_W = \langle \mathcal{P}\varphi, \mathcal{P}u \rangle_X = (\varphi, \overline{\mathcal{P}u})$$

with $\varphi \in C_0^\infty(T)$, where $x = \mathcal{P}\bar{u}$ with $u \in W$ represent all Schwartz distributions $x \in X$ — see (3.4). It shows that the orthogonal complement to $\mathcal{P}C_0^\infty(T)$ in X consists of all $x \in X$ with supports $\operatorname{supp} x \subseteq T^c$ in the complement $T^c = R^d \setminus T$ of the region T, that is

$$(\varphi, x) = 0, \qquad \varphi \in C_0^\infty(T).$$

Obviously, for $x \in X(T)$ with supports $\operatorname{supp} x \subseteq [T]$ in the closure of T it means that $\operatorname{supp} x \subseteq \Gamma = \partial T$. And considering the Sobolev space $X = \overset{\circ}{W}{}_2^{-p}(R^d)$ with the equivalent norm $\|x\|_{-p}$, $x \in X$, we are getting (4.8) as the *direct sum.* ■

Note that we have essentially enlarged variety of differential operators $\mathcal{P}$ which can be applied in (4.8). Namely, we can take any *nondegenerate elliptic operator* $\mathcal{P} \geq 0$,

$$(\varphi, \mathcal{P}\bar{\varphi}) \asymp \|\varphi\|_p^2, \qquad \varphi \in C_0^\infty(T_0),$$

in some region $T_0 \supseteq [T]$. To explain, we can relate the considered above scheme to the Sobolev spaces

$$W = [C_0^\infty(T_0)] = \overset{\circ}{W}{}_2^{p}(T_0), \qquad X = [C_0^\infty(T_0)] = \overset{\circ}{W}{}_2^{-p}(T_0)$$

with the corresponding norms

$$\|u\|_W = (u, \mathcal{P}\bar{u})^{1/2} \asymp \|u\|_p, \qquad u \in W,$$

and

$$\|x\|_X = \sup_{\|u\|_W \leq 1} |(x,u)| \asymp \sup_{\|u\|_p \leq 1} |(x,u)| = \|x\|_{-p}$$

— cf. (3.25); this let us treat our test function space $X(T)$ as the subspace

$$X(T) = [C_0^\infty(T)] \subseteq X = \overset{\circ}{W}{}_2^{-p}(T_0). \tag{4.8$'$}$$

4.3. Boundary test functions

Considering the test functions space

$$X(T) = [C_0^\infty(T)] \subseteq X = \overset{\circ}{W}{}_2^{-p}(R^d)$$

in the region $T \subseteq R^d$, we introduced the *boundary test functions* as Schwartz distributions in X with supports on the boundary $\Gamma = \partial T$. We are going to give their characterization in a case of the piecewise regular boundary Γ.

Let us consider the regular boundary $\Gamma \subseteq \cup O_\alpha$ with the finite map of local coordinates

$$O_\alpha \ni t = \tau_\alpha(s,r), \qquad (s,r) \in R^{d-1} \times R^1,$$

which let us to determine the nontangent derivatives

$$\varphi^{(k)}(t) = \frac{\partial^k}{\partial l^k}\varphi(t) = \frac{\partial^k}{\partial r^k}\varphi(\tau_\alpha(s,0)), \qquad t \in \Gamma \quad (k = 0, \dots, p-1)$$

— see (4.4), and to introduce the Schwartz distributions

$$x \times \delta^{(k)} = (x \times \delta^{(k)}, \varphi) = (-1)^k \int_\Gamma x(t)\varphi^{(k)}(t)\, dt, \qquad \varphi \in D = C_0^\infty(R^d), \tag{4.9}$$

associated with the corresponding weight functions $x = x(t)$, $t \in \Gamma$, on the boundary Γ. Taking a totality of $x \in C_0^p(\Gamma)$, say, we see that we can determine the derivatives $\varphi^{(k)}(t)$, $t \in \Gamma$, of $\varphi \in C_0^\infty(R^d)$ by means of $x \times \delta^{(k)}$ as the certain generalized test functions with supports on Γ. ■

At first, let us consider Γ which in total is the very hyperplane

$$\Gamma = R^{d-1} \subseteq R^d$$

described by the proper coordinates as

$$\Gamma = \{t = (s,r)\colon\ s \in R^{d-1},\ r = 0\}$$

— cf. (4.3); it serves as the boundary for the regions $T = T_- = R^{d-1} \times (-\infty, 0)$ and $T_+ = R^{d-1} \times (0, \infty)$. With

$$t = (s,r), \qquad (s,r) \in R^{d-1} \times R^1$$

we have in (4.9) the Schwartz distributions

$$x \times \delta^{(k)} = (x \times \delta^{(k)}, \varphi) =$$

$$= (-1)^k \int\limits_{R^{d-1}} x(s)\varphi^k(s,0)\, ds, \qquad \varphi \in D = C_0^\infty(R^d), \tag{4.9$'$}$$

which in its totality with the weight functions $x \in C_0^p(\Gamma)$, say, determine the normal derivatives

$$\varphi^{(k)}(s,0) = \frac{\partial^k}{\partial r^k}\varphi(s,r)\Big|_{r=0}, \qquad s \in R^{d-1},$$

on the boundary $\Gamma = R^{d-1}$.

Let us turn on the Fourier transform

$$\widetilde{x \times \delta^{(k)}} = \frac{(-1)^k}{(2\pi)^{d/2}} \int\limits_{R^{d-1}} x(s)(i\mu)^k e^{-i\lambda s}\, ds =$$

$$= \frac{(-1)^k}{\sqrt{2\pi}}\tilde{x}(\lambda) \times (i\mu)^k, \qquad (\lambda,\mu) \in R^{d-1} \times R^1,$$

of these Schwartz distributions. Taking $a^{2p} = 1 + |\lambda|^{2p}$, we have

$$\int\limits_{R^d} |\widetilde{x \times \delta^{(k)}}|^2 (a^{2p} + \mu^{2p})^{-1}\, d\lambda\, d\mu =$$

$$= \frac{1}{2\pi} \iint\limits_{R^{d-1}\times R^1} |\tilde{x}(\lambda)|^2 \frac{\mu^{2k}}{a^{2p} + \mu^{2p}}\, d\lambda\, d\mu =$$

$$= \frac{1}{2\pi} \int\limits_{R^{d-1}} |\tilde{x}(\lambda)|^2 \left[\frac{a^{2k+1}}{a^{2p}} \int\limits_{R^1} \frac{(\frac{\mu}{a})^{2k}\, d(\frac{\mu}{a})}{1 + (\frac{\mu}{a})^{2p}}\right] d\lambda =$$

$$= C \int\limits_{R^{d-1}} |\tilde{x}(\lambda)|^2 a^{-2(p-k-1/2)}\, d\lambda < \infty$$

for $k = 0, \ldots, p-1$. This shows that

$$x \times \delta^{(k)} \in X = \overset{\circ}{W}{}_2^{-p}(R^d), \qquad k = 0, \ldots, p-1, \tag{4.10}$$

according to the known Fourier transform $\widetilde{X} = L_{2,1/F}$ as the $L_{2,1/F}$-space with the weight function

$$1/F \asymp (1 + |\lambda|^{2p} + \mu^{2p})^{-1}, \qquad (\lambda,\mu) \in R^{d-1} \times R^1,$$

— see (3.9), (3.10). Moreover, we see that

$$\|x \times \delta^{(k)}\|_{-p} \asymp \|x\|_{-(p-k-1/2)}, \qquad x \in C_0^p(\Gamma), \tag{4.11}$$

and a closure $X^{(k)}(\Gamma)$ of all $x \times \delta^{(k)}$, $x \in C_0^p(\Gamma)$, with the correspondence

$$X^{(k)}(\Gamma) \ni x \times \delta^{(k)} \leftrightarrow x \in \overset{\circ}{W}{}_2^{-(p-k-1/2)}(\Gamma)$$

can be identified by the exact embedding

$$X^{(k)}(\Gamma) = \overset{\circ}{W}{}_2^{-(p-k-1/2)}(\Gamma) \tag{4.12}$$

to the corresponding *Sobolev space* on the boundary $\Gamma = R^{d-1}$ with the norm

$$\|x\|^2_{-(p-k-1/2)} = \int\limits_{R^{d-1}} |\tilde{x}(\lambda)|^2 (1 + |\lambda|^{2(p-k-1/2)})^{-1} \, d\lambda. \quad \blacksquare$$

Application of the boundary test functions $x \times \delta^{(k)}$ to any $u \in W = \overset{\circ}{W}{}_2^p(R^d) = X^*$ with the result

$$(x \times \delta^{(k)}, u) = (-1)^k (x, u^{(k)})$$

determines on the boundary Γ the generalized functions

$$u^{(k)} = (x, u^{(k)}), \qquad x \in C_0^p(\Gamma), \quad k = 0, \ldots, p-1. \tag{4.13}$$

These functions, according to the equivalence (4.11), are continuous with respect to the corresponding norm $\|x\|_{-(p-k-1/2)}$. Hence,

$$u^{(k)} \in \overset{\circ}{W}{}_2^{p-k-1/2}(\Gamma)$$

belong to the corresponding Sobolev space, dual to $\overset{\circ}{W}{}_2^{-(p-k-1/2)}(\Gamma)$, and we have

$$\|u^{(k)}\|_{p-k-1/2} \asymp \sup_{\|x \times \delta^{(k)}\|_{-p} \leq 1} (x \times \delta^{(k)}, u) \leq C\|u\|_p \tag{4.14}$$

for all $u \in \overset{\circ}{W}{}_2^p(R^d)$.

Taking in account the embeddings

$$\overset{\circ}{W}{}_2^{p-k-1/2}(\Gamma) \subseteq L_2(\Gamma) \subseteq \overset{\circ}{W}{}_2^{-(p-k-1/2)}(\Gamma)$$

we can treat $u^{(k)} \in \overset{\circ}{W}{}_2^{p-k-1/2}(\Gamma)$ for all $k = 0, \dots, p-1$ as the pointwise functions $u^{(k)} = u^{(k)}(t)$, $t \in \Gamma$, in the $L_2(\Gamma)$-space on the boundary Γ. How are these functions related to the boundary trace (4.4) of the function $u \in \overset{\circ}{W}{}_2^p(R^d)$ nontangent derivatives?

LEMMA. *The functions* $u^{(k)} = u^{(k)}(t)$, $t = (s, 0) \in \Gamma$, *represent the derivatives boundary trace*

$$u^{(k)}(t) = \frac{\partial^k}{\partial r^k} u(s, r)\Big|_{r=0}, \qquad k = 0, \dots, p-1.$$

PROOF. Taking delta-approximations $w_n \in C_0^\infty(-\infty, 0)$ to the delta-function δ_0 at $r = 0$, we have the Fourier transform $\tilde{w}_n(\mu) \to \tilde{\delta}_0 = 1/(2\pi)^{d/2}$ and

$$\|x \times w_n^{(k)} - x \times \delta^{(k)}\|_{-p} \asymp$$

$$\asymp \iint\limits_{R^{d-1} \times R^1} |\tilde{x}(\lambda)|^2 \frac{|\tilde{w}_n(\mu) - \tilde{\delta}_0|^2 \mu^{2k}}{1 + |\lambda|^{2p} + \mu^{2p}} \, d\lambda \, d\mu \to 0, \qquad n \to \infty,$$

for all $x \in L_2(\Gamma)$ on the boundary $\Gamma = R^{d-1}$. Hence, for any $u \in \overset{\circ}{W}{}_2^p(R^d)$ we have

$$(x \times w_n^{(k)}, u) = (-1)^k \int\limits_{R^{d-1}} x(s) u_n^{(k)}(s) \, ds \to$$

$$\to (-1)^k \int\limits_{R^{d-1}} x(s) u^{(k)}(s, 0) \, ds = (x \times \delta^{(k)}, u)$$

with

$$u_n^{(k)}(s) = (-1)^k \int\limits_{R^1} w_n^{(k)}(r) u(s, r) \, dr =$$

$$= \int\limits_{R^1} w_n(r) \frac{\partial^k u(s, r)}{\partial r^k} \, dr \to \frac{\partial^k u(s, r)}{\partial r^k}\Big|_{r=0}$$

for almost all $s \in R^{d-1} = \Gamma$. Here, with $u_n^{(k)}$ as elements of the $L_2(\Gamma)$-space, we have their weak convergence to the function $u^{(k)} = u^{(k)}(t)$, $t = (s, 0) \in \Gamma$, in (4.13) and, in the same time, we have their convergence almost everywhere on Γ to the derivative $\frac{\partial^k u(s,r)}{\partial r^k}\big|_{r=0}$. The proof is over.

What was shown above for $u \in \overset{\circ}{W}{}_2^p(T)$ gives us a good reason to say that (4.13) represents on $\Gamma = \partial T$ the *generalized boundary trace* of the generalized derivatives

$$\frac{\partial^k}{\partial l^k} u = \Big(\varphi, \frac{\partial^k u}{\partial l^k}\Big) = (-1)^k \Big(\frac{\partial^k}{\partial l^k}\varphi, u\Big), \qquad \varphi \in C_0^\infty(T \cap \Gamma^\varepsilon)$$
$$(k = 0, \dots, p-1)$$

along the given vector field $l = l(t)$, $t \in \Gamma$, in the boundary neighborhood $\Gamma^\varepsilon = \cup O_\alpha$. ■

Let us consider the total *boundary test functions space* $X(\Gamma)$ which consists of all Schwartz distributions in $X = \overset{\circ}{W}{}_2^{-p}(R^d)$ with supports on the boundary Γ.

As was shown, all Schwartz distributions of the type (4.9) are in $X(\Gamma)$ for $k = 0, \dots, p-1$, representing the corresponding boundary subspace

$$X^{(k)}(\Gamma) = \{x \times \delta^{(k)}\colon\ x \in \overset{\circ}{W}{}_2^{-(p-k-1/2)}(\Gamma)\}$$

— see (4.12)

THEOREM. *The direct sum representation*

$$X(\Gamma) = \sum_{k=0}^{p-1} \oplus X^{(k)}(\Gamma) \tag{4.15}$$

holds true.

PROOF. The direct sum in the right-hand side of (4.15) means that

$$\Big\| \sum_{k=0}^{p-1} x_k \times \delta^{(k)} \Big\|^2 \asymp \sum_{k=0}^{p-1} \|x_k \times \delta^{(k)}\|_{-p}^2 \tag{4.16}$$

for all $x_k \times \delta^{(k)} \in X^{(k)}(\Gamma)$ with its Fourier transform

$$\widetilde{x_k \times \delta^{(k)}} = \frac{(-1)^k}{\sqrt{2\pi}} \tilde{x}_k(\lambda) \times (i\mu)^k, \qquad (\lambda, \mu) \in R^{d-1} \times R^1.$$

Taking $a^{2p} = 1 + |\lambda|^{2p}$ and $C_k = \tilde{x}_k(\lambda) a^k$, we have

$$\int_{R^1} \Big| \sum_{k=0}^{p-1} C_k (i\mu)^k \Big|^2 \frac{d\mu}{1 + \mu^{2p}} \asymp \sum_{k=0}^{p-1} |C_k|^2 \int_{R^1} |i\mu|^{2k} \frac{d\mu}{1 + \mu^{2p}}$$

for the linearly independent functions $f_k(\mu) = (i\mu)^k$, $\mu \in R^1$. Hence,

$$\left\|\sum_{k=0}^{p-1} x_k \times \delta^{(k)}\right\|_{-p}^2 \asymp \int_{R^{d-1}} \frac{a}{a^{2p}} \int_{R^1} \left|\sum_{k=0}^{p-1} \tilde{x}_k(\lambda) a^k \left(i\frac{\mu}{a}\right)^k\right|^2 \frac{d(\frac{\mu}{a})}{1+(\frac{\mu}{a})^{2p}}\, d\lambda =$$

$$= \int_{R^{d-1}} \frac{a}{a^{2p}} \int_{R^1} \left|\sum_{k=0}^{p-1} C_k (i\mu)^k\right|^2 \frac{d\mu}{1+\mu^{2p}}\, d\lambda \asymp$$

$$\asymp \sum_{k=0}^{p-1} \int_{R^{d-1}} \frac{a}{a^{2p}} |C_k|^2 \int_{R^1} |i\mu|^{2k} \frac{d\mu}{1+\mu^{2p}}\, d\lambda \asymp \sum_{k=0}^{p-1} \|x_k \times \delta^{(k)}\|_{-p}^2.$$

Thus, the equivalence (4.16) holds true.

To prove the equality (4.15), we can show that all boundary test functions $x \times \delta^{(k)}$ of the type (4.9) with $k = 0, \ldots, p-1$ form the complete system in the boundary test functions space $X(\Gamma) \subseteq X = \overset{\circ}{W}{}_2^{-p}(R^d)$. Let us take any $u \in W = \overset{\circ}{W}{}_2^{p}(R^d)$ from the dual space $W = X^*$ such that

$$(x \times \delta^{(k)}, u) = (-1)^k \int_\Gamma x(s) u^{(k)}(s)\, ds = 0, \qquad x \in C_0^p(\Gamma),$$

for $k = 0, \ldots, p-1$. We see that the boundary trace of it's derivatives, defined in (4.4), is zero. Thus, we have

$$u \in \overset{\circ}{W}{}_2^p(R^d \setminus \Gamma) = [C_0^\infty(R^d \setminus \Gamma)]$$

— see (4.5), that is $u = \lim \varphi$ for $\varphi \in C_0^\infty(R^d \setminus \Gamma)$. Hence,

$$(x, u) = \lim_{\varphi \to u} (x, \varphi) = 0$$

for all $x \in X$ with supports $\operatorname{supp} x \subseteq \Gamma$ on the boundary. This shows that

$$x \times \delta^{(k)}, \qquad x \in C_0^\infty(\Gamma), \quad k = 0, \ldots, p-1,$$

form the complete system in $X(\Gamma)$, and it ends the proof. ■

Now let us extend our results on the general boundary $\Gamma = \partial T$. What is supposed to be done? We are to show that the Schwartz distributions (4.9) for $k = 0, \ldots, p-1$ are in our test functions space $X = \overset{\circ}{W}{}_2^{-p}(R^d)$ — see (4.10). Then we are to explain the embedding (4.12) by introduction of the proper Sobolev space

$\overset{\circ}{W}{}_2^{-(p-k-1/2)}(\Gamma)$ with the norm which satisfies the equivalence condition (4.11). Considering the generalized derivatives

$$\frac{\partial^k}{\partial l^k}u = \left(\varphi, \frac{\partial^k}{\partial l^k}u\right) = (-1)^k\left(\frac{\partial^k}{\partial l^k}\varphi, u\right), \qquad \varphi \in C_0^\infty(T \cap \Gamma^\varepsilon)$$

along the given vector field in the boundary neighborhood $\Gamma^\varepsilon = \cup O_\alpha$, we are to take in account that (4.13) with $k = 0, \dots, p-1$ represents their *generalized boundary trace.* Finally, we are to verify the direct sum representation (4.15). ■

The main point is that the regular boundary Γ *locally* in the proper local coordinates, can be treated similar to the hyperplane $\Gamma = R^{d-1}$ — see (4.3). Namely, let us take just a part $\Gamma_\alpha = \Gamma \cap O_\alpha$ in a local neighborhood with the corresponding coordinates

$$t = \tau_\alpha(s, r), \qquad (s, r) \in R^{d-1} \times R^1,$$

of the type (4.2). For the Schwartz distributions (4.9) with weight functions $x = x(t)$, $t \in \Gamma$, localized on Γ_α, we have

$$(x \times \delta^{(k)}, \varphi) = (-1)^k \int_\Gamma x(t)\varphi^{(k)}(t)\,dt = (-1)^k \int_{R^{d-1}} x_\alpha(s)(\tau_\alpha\varphi)^{(k)}\,ds =$$

$$= (\tau_\alpha^*(x_\alpha \times \delta^{(k)}), \varphi), \qquad \varphi \in D = C_0^\infty(R^d),$$

with

$$x_\alpha(s) = x(\tau_\alpha(s, 0))\left|\frac{dt}{ds}\right|, \qquad s \in R^{d-1}.$$

Hence, according to the isometric property of $\tau^* = \tau_\alpha^*$ related to the transformation of the type (4.2), we have

$$x \times \delta^{(k)} = \tau^*(x_\alpha \times \delta^{(k)}) \in X = \overset{\circ}{W}{}_2^{-p}(R^d)$$

with the norm

$$\|x \times \delta^{(k)}\|_{-p} \asymp \|x_\alpha \times \delta^{(k)}\|_{-p} \asymp \|x_\alpha\|_{-(p-k-1/2)}$$

— see (4.11). Moreover, for the localized weight functions we have the *direct sum*

$$\sum_{k=0}^{p-1} x_k \times \delta^{(k)} = \tau^*\left(\sum_{k=0}^{p-1} x_{k,\alpha} \times \delta^{(k)}\right)$$

with

$$\left\| \sum_{k=0}^{p-1} x_k \times \delta^{(k)} \right\|_{-p}^2 \asymp \left\| \sum_{k=0}^{p-1} x_{k,\alpha} \times \delta^{(k)} \right\|_{-p}^2 \asymp$$

$$\asymp \sum_{k=0}^{p-1} \|x_{k,\alpha} \times \delta^{(k)}\|_{-p}^2 \asymp \sum_{k=0}^{p-1} \|x_k \times \delta^{(k)}\|_{-p}^2$$

as it is indicated in (4.16). Obviously, the general Schwartz distributions (4.9) are in our test functions space $X = \overset{\circ}{W}{}_2^{-p}(R^d)$, since

$$x \times \delta^{(k)} = \sum_{\alpha} w_\alpha x \times \delta^{(k)}$$

for the unite decomposition

$$1 = \sum_{\alpha} w_\alpha, \qquad w_\alpha \in C_0^p(O_\alpha), \tag{4.17}$$

on the boundary $\Gamma = \cup O_\alpha$ with all weight functions $w_\alpha x$ localized on the corresponding boundary part $\Gamma_\alpha = \Gamma \cap O_\alpha$. The decomposition

$$x \times \delta^{(k)} = \sum_{\alpha} w_\alpha x \times \delta^{(k)} = \sum_{\alpha} \tau_\alpha^*(x_\alpha \times \delta^{(k)}) \tag{4.18}$$

with the localized weight functions

$$x_\alpha = x_\alpha(s) = w_\alpha x(\tau(s,0))\left|\frac{dt}{ds}\right|, \qquad s \in R^{d-1},$$

will be the main key for the further developments. ■

As seen above, the equivalence (4.16) holds true for localized weight functions. Let us verify that it holds true in general, that is we have

$$\left\| \sum_{k=0}^{m} x_k \times \delta^{(k)} \right\|_{-p}^2 \asymp \left\| \sum_{k=0}^{m-1} x_k \times \delta^{(k)} \right\|_{-p}^2 + \|x_m \times \delta^{(m)}\|_{-p}^2$$

for all $m = 0, \ldots, p-1$. Suppose, it is not so. Then for some $x_m \times \delta^{(m)}$, $\|x_m \times \delta^{(m)}\| \geq c > 0$, there are $\sum_{k=0}^{m} x_k \times \delta^{(k)} \to 0$. The multiplication by $w \in C_b^p(R^d)$ is bounded in the Sobolev space $X = \overset{\circ}{W}{}_2^{-p}(R^d)$, and it gives

$$w(x \times \delta^{(k)}) = \sum_{j=0}^{k} \binom{k}{j} (w^{(k-j)} x \times \delta^{(j)}) \tag{4.19}$$

according to the known Leibnitz formula for differentiation. Therefore, with $\sum_{k=0}^{m} x_k \times \delta^{(k)} \to 0$ we should have

$$w_\alpha \sum_{k=0}^{m} x_k \times \delta^{(k)} = \sum_{k=0}^{m} y_k \times \delta^{(k)} \to 0$$

for all components w_α in the unite decomposition (4.17). Here $y_m = w_\alpha x_m$ and all weight functions y_k, $k = 0, \dots, m$, are localized on Γ_α. Hence, in the direct sum $\sum_{k=0}^{m} y_k \times \delta^{(k)} \to 0$ we should have

$$\|y_m \times \delta^{(m)}\|_{-p} = \|w_\alpha x \times \delta^{(m)}\|_{-p} \to 0$$

which contradicts to the property

$$0 < c \le \|x_m \times \delta^{(m)}\|_{-p} \le \sum_\alpha \|w_\alpha x_m \times \delta^{(m)}\|_{-p}$$

Thus, *the equivalence* (4.16) *holds true in general.* ■

The direct sum in the Leibnitz type formula (4.19) gives

$$\|w(x \times \delta^{(k)})\|_{-p}^2 \asymp \sum_{j=0}^{k} \|w^{(k-j)} x \times \delta^{(j)}\|_{-p}^2,$$

so here in the right-hand side for $j = k$ we have

$$\|wx \times \delta^{(k)}\|_{-p} \le C\|w(x \times \delta^{(k)})\|_{-p},$$

and it shows, in particular, that the *multiplication by* $w \in C_b^p(R^d)$ *of weight functions in* (4.9) *is a bounded operator on the boundary subspace* $X^{(k)}(\Gamma)$.

Recall, $X^{(k)}(\Gamma)$ represents the closure of all test functions (4.9) with the given $k = 0, \dots, p-1$, and the verified equivalence (4.16) means that we have the *direct sum*

$$\sum_{k=0}^{p-1} \oplus X^{(k)}(\Gamma). \quad ■$$

Now let us turn on the exact embedding (4.12). The very *Sobolev space* $\overset{\circ}{W}{}_2^{-(p-k-1/2)}(\Gamma)$ on the general boundary Γ we define as a closure of all $x \in C_0^p(\Gamma)$, say, with respect to the certain norm $\|x\|_{-(p-k-1/2)}$ which can be given as follows. According to the decomposition (4.17), we represent $x = x(t)$, $t \in \Gamma$, as

$$x(t) = \sum_\alpha w_\alpha x(t) = \sum_\alpha x_\alpha(s)\left|\frac{ds}{dt}\right|, \qquad s \in R^{d-1},$$

and *define*

$$\|x\|^2_{-(p-k-1/2)} \asymp \sum_\alpha \|x_\alpha\|^2_{-(p-k-1/2)} \tag{4.20}$$

with the known norm for the functions $x_\alpha \in \overset{\circ}{W}{}_2^{-(p-k-1/2)}(R^{d-1})$. Having in mind the isometric property of the τ^*_α-transformators and the boundness of the w_α-multiplicators, from the decomposition (4.18) we obtain

$$\sum_\alpha \|w_\alpha x \times \delta^{(k)}\|_{-p} = \sum_\alpha \|\tau^*_\alpha(x_\alpha \times \delta^{(k)})\|_{-p} \asymp \sum_\alpha \|x_\alpha \times \delta^{(k)}\|_{-p} \asymp$$

$$\asymp \sum_\alpha \|x_\alpha\|_{-(p-k-1/2)} \asymp \|x\|_{-(p-k-1/2)},$$

$$\|x \times \delta^{(k)}\|_{-p} \le \sum_\alpha \|w_\alpha x \times \delta^{(k)}\|_{-p} \le C\|x \times \delta^{(k)}\|_{-p},$$

so we have

$$\|x \times \delta^{(k)}\|_{-p} \asymp \|x\|_{-(p-k-1/2)}$$

— cf. (4.11). It proves the exact embedding (4.12) and also shows that the Sobolev space $\overset{\circ}{W}{}_2^{-(p-k-1/2)}(\Gamma)$ with the norm (4.20) does not depend on the particular unite decomposition (4.17) which was the very key in the scheme applied. Note, we have the equivalence

$$\|x\|^2_{-(p-k-1/2)} \asymp \int_{R^d} \left| \int_\Gamma x(t)(il\lambda)^k e^{-i\lambda t}\, dt \right|^2 (1+|\lambda|^{2p})^{-1}\, d\lambda \tag{4.20$'$}$$

where the right-hand side represents the norm

$$\|x \times \delta^{(k)}\|^2_{-p} = \int_{R^d} |\widetilde{x \times \delta^{(k)}}|^2 (1+|\lambda|^{2p})^{-1}\, d\lambda$$

for $x \times \delta^{(k)} \in X = \overset{\circ}{W}{}_2^{-p}(R^d)$ with $x \in C_0^p(\Gamma)$, say, and the kth derivatives along the given vector field $l = l(t)$, $t \in \Gamma$, — see (4.3), (4.3)$'$.

Obviously, for all $k = 0, \dots, p-1$ we have

$$\|x\|_{-(p-k-1/2)} \asymp \sum_\alpha \|x_\alpha\|_{-(p-k-1/2)} \le$$

$$\le \sum_\alpha \|x_\alpha\|_{L_2(R^{d-1})} = \sum_\alpha \|w_\alpha x\|_{L_2(\Gamma)} \le C\|x\|_{L_2(\Gamma)},$$

so the embeddings

$$\overset{\circ}{W}{}_2^{p-k-1/2}(\Gamma) \subseteq L_2(\Gamma) \subseteq \overset{\circ}{W}{}_2^{-(p-k-1/2)}(\Gamma)$$

for $\overset{\circ}{W}{}_2^{-(p-k-1/2)}(\Gamma)$ and it's dual space $\overset{\circ}{W}{}_2^{p-k-1/2}(\Gamma)$ holds true. And now we can go along the very same way as before, to obtain the rest of our results (including the direct sum representation (4.15)) for the general regular boundary Γ. ■

Let us turn on the piecewise regular boundary $\Gamma = \partial T$ with singularities of the rectangular type — see (4.7).

As before, we can conclude that the *Schwartz distributions* (4.9) *form complete system in the boundary test functions space* $X(\Gamma)$.

As before, taking any regular part Γ_{reg} of the boundary Γ, we obtain that *the exact embedding*

$$X^{(k)}(\Gamma_{\text{reg}}) = \overset{\circ}{W}{}_2^{-(p-k-1/2)}(\Gamma_{\text{reg}})$$

holds true for $X^{(k)}(\Gamma_{\text{reg}})$ representing the closure of the boundary test functions $x \times \delta^{(k)}$ of the type (4.9) with the weight functions $x = x(t)$, $t \in \Gamma$, having supports $\operatorname{supp} x \subseteq \Gamma_{\text{reg}}$. And, as before, these $X^{(k)}(\Gamma_{\text{reg}})$ *form the direct sum*

$$\sum_{k=0}^{p-1} \oplus X^{(k)}(\Gamma_{\text{reg}}),$$

though, in general, we have no more the direct sum representation (4.15) for the total boundary Γ.

For example, let us consider the rectangular $T = R^2_+(t_1, t_2)$ on the R^2-plane with the boundary $\Gamma = \Gamma_1 \cup \Gamma_2$ having the regular parts

$$\Gamma_1 = \{t = (t_1, t_2)\colon\ t_1 > 0,\ t_2 = 0\},$$

$$\Gamma_2 = \{t = (t_1, t_2)\colon\ t_2 > 0,\ t_1 = 0\}.$$

Considering the boundary test functions $x_0(t_1) \times \delta^{(0)}(t_2)$ and $x_1(t_2) \times \delta^{(1)}(t_1)$ in the Sobolev space $X = \overset{\circ}{W}{}_2^{-p}(R^2)$, $p \geq 2$, with the Fourier transform

$$\widetilde{x_0 \times \delta^{(0)}} = \frac{1}{\sqrt{2\pi}} \tilde{x}_0(\lambda_1),$$

$$\widetilde{x_1 \times \delta^{(1)}} = \frac{1}{\sqrt{2\pi}} \tilde{x}_1(\lambda_2) \times (i\lambda_1), \qquad (\lambda_1, \lambda_2) \in R^2,$$

we can take the weight functions $x_0(t_1) \to \delta^{(1)}(t_1)$, $x_1(t_2) \to \delta^{(0)}(t_2)$ approximating the delta-function derivative $\delta^{(1)}$ and the delta-function $\delta^{(0)}$. Then

$$\tilde{x}_0(\lambda_1) \to \frac{1}{\sqrt{2\pi}}(i\lambda_1), \qquad \tilde{x}_1(\lambda_2) \to \frac{1}{\sqrt{2\pi}}$$

and

$$x_0 \times \delta^{(0)} - x_1 \times \delta^{(1)} \to 0.$$

Note that in the regular case the direct sum representation (4.15) with the exact embeddings (4.12) alltogether give a quite explicit characterization of the boundary test functions space $X(\Gamma)$. In general, we have $X(\Gamma)$ as the closure of all elements $\sum_{k=0}^{p-1} x_k \times \delta^{(k)}$, and it can be identified by mapping

$$\sum_{k=0}^{p-1} x_k \times \delta^{(k)} \leftrightarrow x = (x_1, \dots, x_{p-1})$$

with the corresponding vector weight-functions space $X(\Gamma)$ on the boundary, generated by $x = x(t)$, $t \in \Gamma$, with the norm

$$\|x\|^2_{X(\Gamma)} \asymp \int_{R^d} \left| \int_\Gamma \sum_{k=0}^{p-1} x_k(t)(i\lambda l)^k e^{-i\lambda t}\, dt \right|^2 (1 + |\lambda|^{2p})^{-1}\, d\lambda, \qquad (4.21)$$

where the right-hand side is actually represented by the norm $\| \sum_{k=0}^{p-1} x_k \times \delta^{(k)} \|^2_{-p}$ in $X = \overset{\circ}{W}{}_2^{-p}(R^d)$. Recall, that

$$\|x\|^2_{X(\Gamma)} \asymp \sum_{k=0}^{p-1} \|x_k\|^2_{-(p-k-1/2)}, \qquad \operatorname{supp} x \subseteq \Gamma_{\text{reg}},$$

on any regular part Γ_{reg} of the boundary Γ. ■

4.4. Anisotropic Sobolev spaces

In the framework of 4.1–4.3 we are going to consider

$$W = \overset{\circ}{W}{}_2^{(p_1,p_2)}(R^d), \qquad X = \overset{\circ}{W}{}_2^{-(p_1,p_2)}(R^d)$$

with $p_1 \neq p_2$ as typical examples of anisotropic Sobolev spaces (for their general definition see p. 24).

By the corresponding pointwise characterization we have $u \in \overset{\circ}{W}{}_2^{(p_1,p_2)}(R^{d_1} \times R^{d_2})$ as the functions

$$u = u(t), \qquad t = (t_1, t_2) \in R^{d_1} \times R^{d_2},$$

having derivatives $f = \partial^k u$ of the type (4.1) with $0 \leq |k| \leq p_1 - 1$ for $t_1 \in R^{d_1}$, $0 \leq |k| \leq p_2 - 1$ for $t_2 \in R^{d_2}$.

We are to deal with certain properties related to a boundary $\Gamma = \partial T$ of a region $T \subseteq R^d$. In general, considering a direct product $T = T_1 \times T_2$ (with $T_1 \subseteq R^{d_1}$, $T_2 \subseteq R^{d_2}$), we can apply appropriate coordinate transformation

$$t_1 = \tau_1(\hat{t}_1), \qquad t_2 = \tau_2(\hat{t}_2)$$

of the type (4.2) which gives the corresponding one-to-one equivalently isometric mapping τ in W and it's dual mapping τ^* in $X = W^*$.

For a further convinience we just consider a case of the Sobolev spaces

$$W = \overset{\circ}{W}{}_2^{(p,1)}(R^{d-1} \times R^1), \qquad X = \overset{\circ}{W}{}_2^{-(p,1)}(R^{d-1} \times R^1)$$

with $p_1 = p$, $p_2 = 1$ and $d_1 = d - 1$, $d_2 = 1$.

Let $T = G \times I$ be a cylinder with a region $G \subseteq R^{d-1}$ and an interval $I = (a, b) \subseteq R^1$. Let ∂G be a regular boundary of $G \subseteq R^{d-1}$ with a nontangent vector field $l = l(s)$, $s \in \partial G$, given as it was in our general definition of the regular boundary — see (4.3). We introduce the corresponding vector field

$$l = l(s) \in R^{d-1} \times \{r\}, \qquad s \in \partial G, \qquad r \in I,$$

on the side part $\partial G \times I$ of the cylinder boundary

$$\Gamma = (\partial G \times I) \cup (G \times \partial I).$$

Any $u \in \overset{\circ}{W}{}_2^{(p,1)}(R^{d-1} \times R^1)$ for almost every $t_2 = r \in R^1$ as function of $t_1 \in R^{d-1}$ represents $u_r \in \overset{\circ}{W}{}_2^{p}(R^{d-1})$, since

$$\|u\|_{(p,1)}^2 \asymp \int_{R^1} \|u_r\|_p^2 \, dr + \left\| \frac{\partial u}{\partial r} \right\|_{L_2(R^d)}^2.$$

Hence, we can introduce the corresponding *boundary trace*

$$u^{(k)}(t) = u^{(k)}(s,r), \qquad t=(s,r)\in \partial G\times I \quad (k=0,\dots,p-1) \tag{4.22}$$

on the side boundary where $u^{(k)}(s,r)$, $s\in\partial G$ $(k=0,\dots,p-1)$ represent the known boundary trace for $u_r \in \overset{\circ}{W}{}_2^p(R^{d-1})$ — cf. (4.4). And having $u \in \overset{\circ}{W}{}_2^{(p,1)}(R^{d-1}\times R^1)$ for almost every $s\in R^{d-1}$ absolutely continuous with respect to $r\in R^1$, we can introduce the *boundary trace*

$$u^{(0)}(t) = u(s,r), \qquad t=(s,r)\in G\times\partial I \tag{4.22$'$}$$

on the bottom/top part $G\times\partial I$ of the boundary $\Gamma=\partial T$ (with $r=a$ on the bottom and $r=b$ on the top). Similar to what was obtained in 4.1, in a case of the zero boundary trace (4.22), (4.22)$'$ we have $u\in\overset{\circ}{W}{}_2^{(p,1)}(R^d\setminus\Gamma)$, and $u\in\overset{\circ}{W}{}_2^{(p,1)}(T)$ if $\operatorname{supp} u\subseteq[T]$.

Let us turn on $X=\overset{\circ}{W}{}_2^{-(p,1)}(R^d)$ as the certain test functions space which consists of all Schwartz distributions $x\in D^*$ with

$$\|x\|^2_{-(p,1)} = \int\limits_{R^{d-1}\times R^1} |\tilde{x}(\lambda_1,\lambda_2)|^2(1+|\lambda_1|^{2p}+|\lambda_2|^2)^{-1}\,d\lambda_1\,d\lambda_2<\infty$$

and appears as the closure of $D=C_0^\infty(R^d)$. The corresponding test functions space $X(T)=[C_0^\infty(T)]$ in the region $T\subseteq R^d$ consists of all Schwartz distributions $x\in X$ with supports $\operatorname{supp} x\subseteq[T]$. (Note, the same test functions space $X(T)=[C_0^\infty(T)]$ appears as the closure of $C_0^\infty(T)$ in $X=\overset{\circ}{W}{}_2^{-(p,1)}(T_0)$ as dual space to $W=\overset{\circ}{W}{}_2^{(p,1)}(T_0)$ in a region $T_0\supseteq[T]$ — cf. (3.25).)

Hence, the boundary test functions space

$$X(\Gamma)=\{x\in X(T)\colon \operatorname{supp} x\subseteq\Gamma\}$$

consists of all $x\in X$, $\operatorname{supp} x\subseteq\Gamma$.

The direct sum representation

$$X(T)=[\mathcal{P}C_0^\infty(T)]\oplus X(\Gamma)$$

holds true for any differential operator $\mathcal{P}\geq 0$,

$$(\varphi,\mathcal{P}\varphi)^{1/2}\asymp\|\varphi\|_{(p,1)}, \qquad \varphi\in C_0^\infty(T_0),$$

associated with the norm in $W = \overset{\circ}{W}{}_2^{(p,1)}(T_0)$, $T_0 \supseteq [T]$ — cf. (4.8), (4.8)$'$. ■

The boundary test functions space $X(\Gamma)$ is generated by Schwartz distributions $x \times \delta^{(k)}$ of the known type (4.9) with the weight-functions x, $\operatorname{supp} x \subseteq \partial G \times I$, on the side part of the boundary Γ for $k = 0, \dots, p-1$ and with the weight functions x, $\operatorname{supp} x \subseteq G \times \partial I$, on it's bottom/top part for $k = 0$. Similar to (4.13), we can introduce the *generalized boundary trace*

$$u^{(k)} = (x, u^{(k)}) = (x \times \delta^{(k)}, u), \qquad x \in C_0^p(\partial G \times I), \quad k = 0, \dots, p-1, \tag{4.23}$$

of the corresponding generalized derivatives along the given vector field on the side boundary and the *generalized boundary trace*

$$u^{(0)} = (x, u^0) = (x \times \delta^{(0)}, u), \qquad x \in C_0^p(G \times \partial I) \tag{4.23$'$}$$

of the very function $u \in \overset{\circ}{W}{}_2^{(p,1)}(T)$ on the bottom/top of the cylinder $T = G \times I$.

Taking a region G_{loc} with it's closure $[G_{\text{loc}}] \subseteq G$ and an interval I_{loc}, $[I_{\text{loc}}] \subseteq I = (a, b)$ we can form a regular boundary part as

$$\Gamma_{\text{reg}} = (G_{\text{loc}} \times \{a\}) \cup (\partial G \times I_{\text{loc}}) \cup (G_{\text{loc}} \times \{b\}).$$

For $k = 0, \dots, p-1$ the closure $X^{(k)}(\partial G \times I_{\text{loc}})$ of the boundary test functions $x \times \delta^{(k)}$, $\operatorname{supp} x \subseteq \partial G \times I_{\text{loc}}$, on the side boundary part can be characterized by the exact embedding

$$X^{(k)}(\partial G \times I_{\text{loc}}) = \overset{\circ}{W}{}_2^{-(p-k-1/2,1-k/p-1/(2p))}(\partial G \times I_{\text{loc}})$$

with the correspondence of $x \times \delta^{(k)}$ to their weight functions $x = x(t)$, $t \in \partial G \times I_{\text{loc}}$ — cf. (4.12). In a similar way, the test functions $x \times \delta^{(0)}$ on the bottom/top boundary part, considered separately as $x \times \delta_a^{(0)}$, $\operatorname{supp} x \subseteq G_{\text{loc}} \times \{a\}$, and $x \times \delta_b^{(0)}$, $\operatorname{supp} x \subseteq G_{\text{loc}} \times \{b\}$, with the corresponding delta-functions at the boundary points of $I = (a, b)$ can be characterized by the exact embeddings

$$X^{(0)}(G_{\text{loc}} \times \{a\}) = \overset{\circ}{W}{}_2^{-p/2}(G_{\text{loc}}), \qquad X^{(0)}(G_{\text{loc}} \times \{b\}) = \overset{\circ}{W}{}_2^{-p/2}(G_{\text{loc}}).$$

And we have the direct sum

$$X^{(0)}(G_{\text{loc}} \times \{a\}) \oplus \sum_{k=0}^{p-1} \oplus X^{(k)}(\partial G \times I_{\text{loc}}) \oplus X^{(0)}(G_{\text{loc}} \times \{b\})$$

which (with G_{loc} and I_{loc} approximating G and I) gives the corresponding element-wise approximation of the total boundary test functions space $X(\Gamma)$. Similar to what was suggested for the piecewise regular boundary — see (4.21), $X(\Gamma)$ can be identified with the certain vector weight-functions space on Γ. ■

We conclude with the observation of a certain unexpected phenomenon. To explain it, we consider just a case of dimension $d = 2$ and the boundary being a straight line

$$\Gamma = \{t = (t_1, t_2)\colon\ t_1 = t_2\}.$$

In general, there is no pointwise derivative $\partial u/\partial t_2$ of $u \in \overset{\circ}{W}{}_2^{(p,1)}(R^1 \times R^1)$, so there is no pointwise normal derivatives

$$\frac{\partial^k u}{\partial l^k} = \Big(\frac{1}{\sqrt{2}}\frac{\partial}{\partial t_1} - \frac{\partial}{\partial t_2}\Big)^k u(t), \qquad t \in \Gamma.$$

Nevertheless, for the *generalized* normal derivatives

$$\frac{\partial^k}{\partial l^k}u = \Big(\varphi, \frac{\partial^k}{\partial l^k}u\Big) = (-1)^k\Big(\frac{\partial^k \varphi}{\partial l^k}, u\Big), \qquad \varphi \in C_0^\infty(T),$$

with $k = 0, \ldots, p-1$ we have their *generalized boundary trace*

$$u^{(k)} = (x, u^{(k)}) = (-1)^k (x \times \delta^{(k)}, u), \qquad x \in C_0^p(\Gamma)$$

— cf. (4.13). It can be verified by means of the Fourier transform

$$\widetilde{x \times \delta^{(k)}} = \tilde{x}(\lambda_1 + \lambda_2)[i(\lambda_1 - \lambda_2)]^k, \qquad (\lambda_1, \lambda_2) \in R^1 \times R^1,$$

— namely, we have

$$\int\limits_{R^1 \times R^1} |\widetilde{x \times \delta^{(k)}}|^2 (1 + \lambda_1^{2p} + \lambda_2^2)^{-1}\, d\lambda_1\, d\lambda_2 \le C \int\limits_{R^1} |\widetilde{x(\lambda)}|^2 (1 + \lambda^{2k})\, d\lambda < \infty$$

for all $x \in C_0^p(\Gamma)$. We also see here, that this boundary trace $u^{(k)} = (x, u^{(k)})$, $x \in C_0^p(\Gamma)$, is represented by essentially generalized functions on the boundary Γ which in a scale of Sobolev spaces can be characterized not better than as $u^{(k)} \in \overset{\circ}{W}{}_2^{-k}(\Gamma)$ for $k = 0, \ldots, p-1$. ■

4.5. Dual Sobolev spaces

Dealing with an appropriate test functions space $X(T) = [C_0^\infty(T)]$ in a region $T \subseteq R^d$ we introduce the corresponding dual space $W(T) = X(T)^*$, represented with Schwartz distributions

$$u = (\varphi, u), \qquad \varphi \in C_0^\infty(T),$$

which are continuous with respect to a norm $\|\varphi\|_X$ in $X(T)$.

Let us consider $X(T) \subseteq X$ in the Sobolev space $X = \overset{\circ}{W}{}_2^{-p}(R^d)$ with the norm

$$\|\varphi\|_X \asymp \|\varphi\|_{-p}, \qquad \varphi \in C_0^\infty(T).$$

Any $u \in W(T) = X(T)^*$ can be extended to some $u \in \overset{\circ}{W}{}_2^p(R^d) = X^*$ with the norm

$$\|u\|_p \asymp \sup_{\|\varphi\|_{-p} \leq 1} (\varphi, u) = \|u\|_{W(T)}$$

where

$$\|u\|_p^2 \asymp \sum_{|k| \leq p} \|\partial^k u\|_{L_2(R^d)}^2,$$

and it shows that

$$\sum_{|k| \leq p} \|\partial^k u\|_{L_2(T)}^2 \leq C\|u\|_{W(T)}^2.$$

The condition $\partial^k u \in L_2(T)$, $|k| \leq p$, describes the known *Sobolev space* $W_2^p(T)$ with the norm

$$\|u\|^2 = \sum_{|k| \leq p} \|\partial^k u\|_{L_2(T)}^2,$$

so we have $W(T) \subseteq W_2^p(T)$. Any $u \in W_2^p(T)$, in the region $T \subseteq R^d$ with piece-wise regular boundary, can be extended outside T up to some function $u \in W_2^p(R^d)$ — to see this one can apply approximations by $u \in C_b^p(T)$ with their proper extensions outside T. It shows that the functions $u \in W_2^p(T)$ actually represent elements $u \in W(T) = X(T)^*$, so with the bounded one-to-one mapping

$$I\colon W(T) \ni u \to u \in W_2^p(T)$$

we have $W_2^p(T) = IW(T)$. Hence, the inverse mapping is also bounded,

$$\|u\|^2_{W(T)} \le C \sum_{|k|\le p} \|\partial^k u\|^2_{L_2(T)}. \tag{4.24}$$

Thus, we obtain the following result.

THEOREM. *The Sobolev space*

$$W(T) = W_2^p(T) \tag{4.25}$$

with the norm

$$\|u\|^2_{W(T)} \asymp \sum_{|k|\le p} \|\partial^k u\|^2_{L_2(T)}$$

is dual to the test functions space $X(T) = [C_0^\infty(T)]$ *in* $X = \overset{\circ}{W}{}_2^{-p}(R^d)$.

And it gives a good reason to set

$$X(T) = W_2^{-p}(T) \tag{4.25$'$}$$

for $X(T) = W_2^p(T)^*$. ■

Similar to this, following the anisotropic scheme 4.4, with the example of $W = \overset{\circ}{W}{}_2^{(p,1)}(R^{d-1} \times R^1)$, $X = \overset{\circ}{W}{}_2^{-(p,1)}(R^{d-1} \times R^1)$ in the cylinder type region $T = G \times I \subseteq R^{d-1} \times R^1$, we obtain that the dual space $W(T) = X(T)^*$ to the closure $X(T) = [C_0^\infty(T)]$ of $C_0^\infty(T)$ in $X = \overset{\circ}{W}{}_2^{-(p,1)}(R^{d-1} \times R^1)$ is the Sobolev space

$$W(T) = W_2^{(p,1)}(T) \tag{4.26}$$

with the norm

$$\|u\|^2_{W(T)} \asymp \|\partial u\|^2_{L_2(T)} + \sum_{|k|\le p} \|\partial^k u\|^2_{L_2(T)}$$

where generalized derivatives $\partial^k u$, $|k| \le p$, and $\partial u = \partial u/\partial r$ related to the corresponding variables $t_1 = s \in G \subseteq R^{d-1}$ and $t_2 = r \in I \subseteq R^1$. Accordingly, we set

$$X(T) = W_2^{-(p,1)}(T) \tag{4.26$'$}$$

for $X(T) = W_2^{(p,1)}(T)^*$.

4.6. Hilbert–Schmidt embeddings

Let $[C_0^\infty(T)]_n$ be a closure of $C_0^\infty(T)$ with respect to the norm

$$\|\varphi\|_n^2 \asymp \int\limits_{R^1} |\tilde{\varphi}(\lambda)|^2 (1+|\lambda|^2)^n \, d\lambda, \qquad \varphi \in C_0^\infty(T),$$

in Sobolev space $\overset{\circ}{W}{}_2^n(R^d)$, that is

$$[C_0^\infty(T)]_n = \begin{cases} \overset{\circ}{W}{}_2^n(T), & n \geq 0, \\ W_2^n(T), & n \leq 0. \end{cases}$$

The identity operator on $C_0^\infty(T)$ gives us the embeddings

$$[C_0^\infty(T)]_n \subseteq [C_0^\infty(T)]_m, \qquad n \geq m. \tag{4.27}$$

THEOREM. *For*

$$n - m > \frac{d}{2}, \tag{4.28}$$

in any bounded region $T \subseteq R^d$, the embeddings (4.27) *are of the Hilbert–Schmidt type.*

PROOF. Let us take an open cube $T_0 = (-\pi a, \pi a)^d \supseteq [T]$, assuming $a = 1$ for notations convinience. The norm $\|\varphi\|_{\pm n}$, $n = 0, 1, \ldots$, for $\varphi \in C_0^\infty(T)$ is equivalent to the norm in the corresponding Sobolev space $\overset{\circ}{W}{}_2^{\pm n}(T_0)$, associated with the operator $\mathcal{P} = (1-\Delta)^n$; namely,

$$\|\varphi\|_n = (\varphi, \mathcal{P}\varphi)^{1/2}, \qquad \varphi \in C_0^\infty(T_0),$$

and

$$\|\varphi\|_{-n} \asymp \sup_{\|u\|_n \leq 1} |(\varphi, u)|, \qquad \varphi \in C_0^\infty(T),$$

where sup is over $u \in C_0^\infty(T_0)$ — cf. (3.25), (4.8)$'$.

Let us take in the space $L_2(T_0)$ the Fourier series

$$\varphi(t) = \sum_k \tilde{\varphi}_k e_k(t)$$

with the orthonormal basis of functions

$$e_k(t) = \frac{1}{(2\pi)^{d/2}} e^{ikt}, \qquad t \in T_0,$$

(where $k = (k_1, \dots, k_d)$ runs over the integer-valued d-dimensional lattice) and introduce the operators J^n,

$$J^n\varphi(t) = \sum_k (1 + |k|^2)^{n/2} \tilde{\varphi}_k e_k(t), \qquad n = 0, 1, \dots.$$

The operator J^n is defined on all elements $\varphi \in L_2(T_0)$ such that

$$\sum_k (1 + |k|^2)^n |\tilde{\varphi}_k|^2 < \infty,$$

and all functions $\varphi \in C_0^\infty(T)$ satisfy this condition, according to what is known about their Fourier transform

$$\tilde{\varphi}_k = \frac{1}{(2\pi)^{d/2}} \int \varphi(t) e^{-ikt}\, dt \to 0, \qquad k \to \infty.$$

Obviously,

$$(1 - \Delta)^n \varphi(t) = \sum_k (1 + |k|^2)^n \tilde{\varphi}_k e_k(t)$$

and

$$\|\varphi\|_n^2 = \sum_k (1 + |k|^2)^n |\tilde{\varphi}_k|^2 = \|J^n \varphi\|_{L_2}^2, \qquad n = 0, 1, \dots,$$

so for an orthonormal basis $\{\varphi_j\}$ in the Sobolev space $\overset{\circ}{W}{}_2^n(T_0)$ we have the orthonormal system $\{J^n\varphi_j\}$ in the space $L_2(T_0)$.

Let us introduce the operator $Q = J^{-p}$, $p > d/2$. It is a bounded symmetric operator in $L_2(T_0)$, having elements of the orthonormal basis $\{e_k\}$ as its eigenfunctions with the corresponding eigenvalues $(1 + |k|^2)^{-p/2}$, such that

$$\sum_k \|Qe_k\|^2 = \sum_k (1 + |k|^2)^{-p} < \infty \qquad (p > d/2).$$

Thus, it is the Hilbert–Schmidt operator in $L_2(T_0)$. Hence, with $J^m = QJ^n$, $m = n - p$ $(m \geq 0)$ we have

$$\sum_j \|\varphi_j\|_m^2 = \sum_j \|Q(J^n\varphi_j)\|_{L_2}^2 \leq C < \infty$$

for any orthonormal system $\{\varphi_j\}$ in $\overset{\circ}{W}{}_2^n(T_0)$, so

$$\overset{\circ}{W}{}_2^n(T_0) \subseteq \overset{\circ}{W}{}_2^m(T_0), \qquad n - m > d/2,$$

is the Hilbert–Schmidt embedding, as well as the corresponding embedding (4.27). Consequently, with the same m, n we have the Hilbert–Schmidt embedding

$$\overset{\circ}{W}{}_2^{-m}(T_0) \subseteq \overset{\circ}{W}{}_n^{-n}(T_0), \qquad -m-(-n) > d/2$$

for the dual spaces — of course, it also concerns their subspaces

$$[C_0^\infty(T)]_{-m} \subseteq [C_0^\infty(T)]_{-n}.$$

A rest of m, n (not yet considered) consists of $m = n - p$, $-p < m < 0$. In this case, taking orthonormal system $\{\varphi_j\}$ in $\overset{\circ}{W}{}_2^n(T_0)$ with $\varphi_j \in C_0^\infty(T)$, we have

$$\|\varphi_j\|_m \asymp \sup_{\|\varphi\|_{-m}\le 1} (\varphi, \varphi_j)$$

$$= \sup_{\|J^{-m}\varphi\|_{L_2}\le 1} |(J^{-m}\varphi, Q(J^n\varphi_j))| \le \|Q(J^n\varphi_j)\|_{L_2}$$

where sup is over $\varphi \in C_0^\infty(T_0)$ and $\{J^n\varphi_j\}$ is orthonormal system in $L_2(T_0)$, so

$$\sum_j \|\varphi_j\|_m^2 \le C \sum_j \|Q(J^n\varphi_j)\|_{L_2}^2$$

and by this we complete the proof of Hilbert–Schmidt embedding (4.27) for all $m, n = 0, \pm 1, \ldots$. ■

In addition we note, that the condition (4.28) for the Hilbert–Schmidt embedding (4.27) cannot be improved accoridng to the fact that the series

$$\sum_k (1 + |k|^2)^{-p} = \infty, \qquad p \le d/2.$$

I.5. Generalized random functions and stochastic Sobolev spaces

5.1. Random Schwartz distributions

Let $\xi = (\varphi, \xi)$, $\varphi \in C_0^\infty(T)$, be a generalized random function in a region $T \subseteq R^d$, i.e., it is *meansquare* linear continuous function over the Schwartz space $C_0^\infty(T)$. Then, in any bounded subregion

T_{loc}, $[T_{\mathrm{loc}}] \subseteq T$, it is meansquare continuous with respect to the Sobolev norm

$$\|\varphi\|_n^2 = \sum_{|k|\leq n} \|\partial^k \varphi\|_{L_2}^2$$

for some $n = 0, 1, \ldots$.

Indeed, if it be not so, then for every $n = 0, 1, \ldots$ there should be $\varphi_{kn} \in C_0^\infty(T_{\mathrm{loc}})$, $k = 1, 2, \ldots$, with $\|\varphi_{kn}\|_n \leq 1$ such that

$$\|(\varphi_{kn}, \xi)\|^2 = \mathrm{E}|(\varphi_{kn}, \xi)|^2 \geq k^2,$$

say, and therefore $\|(n^{-1}\varphi_{nn}, \xi)\| \geq 1$ with $\|n^{-1}\varphi_{nn}\|_p \to 0$, $n \to \infty$, for all $p = 0, 1, \ldots$, contrary to the meansquare continuity of (φ, ξ) with respect to convergence $\varphi \to 0$ in the Schwartz space $C_0^\infty(T)$.

Thus, we can treat

$$\xi = (\varphi, \xi), \qquad \varphi \in C_0^\infty(T_{\mathrm{loc}})$$

as the generalized random function on the corresponding Sobolev space

$$\overset{\circ}{W}{}_2^n(T_{\mathrm{loc}}) = [C_0^\infty(T_{\mathrm{loc}})].$$

And here, according to the known Hilbert–Schmidt embedding

$$\overset{\circ}{W}{}_2^n(T_{\mathrm{loc}}) \subseteq \overset{\circ}{W}{}_2^m(T_{\mathrm{loc}}), \qquad n - m > d/2,$$

we can take a proper modification of ξ with all its realizations

$$\xi_\omega = (\varphi, \xi_\omega), \qquad \varphi \in C_0^\infty(T_{\mathrm{loc}}),$$

being Schwartz distributions from the dual Sobolev space $\overset{\circ}{W}{}_2^{-m}(T_{\mathrm{loc}})$, that is

$$\xi \in \overset{\circ}{W}{}_2^{-m}(T_{\mathrm{loc}}) \tag{5.1}$$

— cf. (2.1), (2.2).

Let us consider $T_{\mathrm{loc}} = T_k$ $(k = 0, 1, \ldots)$ such that $T_{k-1} \subseteq T_k$,

$$\bigcup_k T_k = T,$$

with the corresponding $m = m_k$ increasing. In every subregion T_k we can take a proper modification $\xi = \xi_k$, say, with all its realizations from the Sobolev space $\overset{\circ}{W}{}_2^{-m_k}(T_k)$, that is

$$\xi_k \in \overset{\circ}{W}{}_2^{-m_k}(T_k).$$

How to stick these ξ_k $(k = 0, 1, \ldots)$ one with another to obtain, as a result, a modification of ξ with all its realizations being Schwartz distributions in the total region $T = \cup T_k$?

We suggest to do it in a quite explicit way by taking certain modifications $\tilde{\xi}_k$ of the given ξ_k as

$$\tilde{\xi}_0 = \xi_0, \quad \tilde{\xi}_k = \tilde{\xi}_{k-1} + \eta_k \qquad (k = 1, 2, \ldots)$$

with appropriate η_k — it let us determine the modification $\tilde{\xi}$ of ξ, we are looking for, as

$$\tilde{\xi} = \lim_{k\to\infty} \tilde{\xi}_k = \xi_0 + \sum_{k=1}^{\infty} \eta_k. \tag{5.2}$$

Namely, according to the known Riesz representation $X = \mathcal{P}W$ with

$$W = \overset{\circ}{W}{}_2^{m_k}(R^d), \qquad X = \overset{\circ}{W}{}_2^{-m_k}(R^d)$$

and $\mathcal{P} = (1 - \Delta)^{m_k}$ — see (3.3), for the subspace

$$\overset{\circ}{W}{}_2^{m_k}(T_{k-1}) = [C_0^\infty(T_{k-1})] \subseteq [C_0^\infty(T_k)] = \overset{\circ}{W}{}_2^{m_k}(T_k)$$

we have

$$\overset{\circ}{W}{}_2^{-m_k}(T_{k-1}) = \mathcal{P}\, \overset{\circ}{W}{}_2^{m_k}(T_{k-1}) \subseteq \mathcal{P}\, \overset{\circ}{W}{}_2^{m_k}(T_k) = \overset{\circ}{W}{}_2^{-m_k}(T_k).$$

In this way, for any realization $\xi_k \in \overset{\circ}{W}{}_2^{-m_k}(T_k)$, we can take a projection $\hat{\xi}_{k-1}$ of this element ξ_k in the Hilbert space $X \supseteq \overset{\circ}{W}{}_2^{-m_k}(T_k)$ onto the subspace $\overset{\circ}{W}{}_2^{-m_k}(T_{k-1}) = \mathcal{P}\, \overset{\circ}{W}{}_2^{m_k}(T_{k-1})$. According to the dual Riesz formula — see (3.4), for $\eta_k = \xi_k - \hat{\xi}_{k-1}$ we have

$$(\varphi, \eta_k) = \langle \mathcal{P}\varphi, \eta_k \rangle_X = 0, \qquad \varphi \in C_0^\infty(T_{k-1}).$$

Hence, with the probability 1 we have

$$\begin{array}{ccccc} (\varphi, \xi) & = & (\varphi, \xi_k) & = & (\varphi, \hat{\xi}_{k-1}) \\ \| & & & & \\ (\varphi, \tilde{\xi}_{k-1}) & & & & \end{array}$$

where the modification $\tilde{\xi}_{k-1} \in \overset{\circ}{W}{}^{-m_{k-1}}(T_{k-1})$ to be defined already. Now we recall the embedding

$$\overset{\circ}{W}{}_2^{-m_{k-1}}(T_{k-1}) \subseteq \overset{\circ}{W}{}_2^{-m_k}(T_{k-1})$$

which let us treat $\tilde{\xi}_{k-1} \in \overset{\circ}{W}{}_2^{-m_{k-1}}(T_{k-1})$ as element of $\overset{\circ}{W}{}_2^{-m_k}(T_{k-1})$ and take it instead of the element $\hat{\xi}_{k-1}$, to form

$$\tilde{\xi}_k = \tilde{\xi}_{k-1} + \eta_k \in \overset{\circ}{W}{}_2^{-m_k}(T_k).$$

Obviously, the formula (5.2) determines the modification $\tilde{\xi}$ of ξ such that

$$(\varphi, \tilde{\xi}) = (\varphi, \tilde{\xi}_k), \qquad \varphi \in C_0^\infty(T_{k-1}), \qquad k = 1, 2, \ldots, \tag{5.2$'$}$$

— recall, that $\cup_k T_{k-1} = T$.

With the reference on (5.2), (5.2)$'$ we obtain, in particular, the following result.

THEOREM. *Any generalized random function can be treated as random Schwartz distribution.* ■

5.2. Stochastic Sobolev spaces

Let ξ be a generalized random function in a region $T \subseteq R^d$. Depending on its "smoothness," we can try to apply, alltogether with $\varphi \in C_0^\infty(T)$, some other test functions, aiming to learn something more than it is given just by means of (φ, ξ), $\varphi \in C_0^\infty(T)$.

For example, in a case of the generalized random function

$$\xi = (\varphi, \xi), \qquad \varphi \in C_0^\infty(T),$$

meansquare continuous with respect to the norm

$$\|\varphi\|_{-p}^2 \asymp \int_{R^d} |\tilde{\varphi}(\lambda)|^2 (1 + |\lambda|^{2p})^{-1} \, d\lambda \tag{5.3}$$

in the Sobolev space $X = \overset{\circ}{W}{}_2^{-p}(R^d)$, we can test ξ by all generalized test functions $x \in X(T) = [C_0^\infty(T)]$ in X with the corresponding result

$$(x, \xi) = \lim_{\varphi \to x} (\varphi, \xi).$$

It gives us ξ as the generalized random function

$$\xi = (x, \xi), \qquad x \in X(T),$$

on the test functions space $X(T) = W_2^{-p}(T)$. This particular class of generalized random functions represents what we suggest to introduce as the *stochastic Sobolev space* $\mathbf{W}_2^p(T)$, equiped with its elements norm

$$\|\xi\|_p = \sup_{\|\varphi\|_{-p}\le 1} (\mathrm{E}|(\varphi,\xi)|^2)^{1/2}. \tag{5.3)'}$$

Obviously, all deterministic $\xi = u \in \mathbf{W}_2^p(T)$ constitute the Sobolev space $W(T) = W_2^p(T)$, dual to $X(T) = W_2^{-p}(T)$ — see (4.25), (4.25)'.

In a similar way anisotropic *stochastic Sobolev spaces* can be introduced. For example, the stochastic Sobolev space $\mathbf{W}_2^{(p,1)}(T)$ in the cylinder type region $T = G \times I \subseteq R^{d-1} \times R^1$ consists of all generalized random functions $\xi = (\varphi,\xi)$, $\varphi \in C_0^\infty(T)$, meansquare continuous with respect to

$$\|\varphi\|^2_{-(p,1)} \asymp \int\limits_{R^{d-1}\times R^1} |\tilde\varphi(\lambda_1,\lambda_2)|^2(1+|\lambda_1|^{2p}+\lambda_2^2)^{-1}\, d\lambda_1\, d\lambda_2 \tag{5.4}$$

with the corresponding norm in $\mathbf{W}^{(p,1)}(T)$ as

$$\|\xi\|_{(p,1)} = \sup_{\|\varphi\|_{-(p,1)}\le 1} (\mathrm{E}|(\varphi,\xi)|^2)^{1/2}. \tag{5.4)'}$$

All deterministic $\xi = u \in \mathbf{W}_2^{(p,1)}(T)$ form the Sobolev space $W(T) = W_2^{(p,1)}(T)$, dual to $X(T) = W_2^{-(p,1)}(T)$ — see (4.26). ■

As one of the most known random fields models, let us turn on the *homogeneous random field*

$$\xi = (\varphi,\xi), \qquad \varphi \in C_0^\infty(R^d),$$

having the zero mean and

$$\|(\varphi,\xi)\|^2 = \mathrm{E}|(\varphi,\xi)|^2 = \int\limits_{R^d} |\tilde\varphi(\lambda)|^2 f(\lambda)\, d\lambda$$

with the *spectral density*

$$f(\lambda) \asymp (1+|\lambda|^{2p})^{-1}, \qquad \lambda \in R^d. \tag{5.5}$$

We see that it is a representative from the stochastic Sobolev space $\mathbf{W}_2^p(R^d)$ — cf. (5.3).

This example shows quite a drastic difference with what we know about *deterministic* Sobolev spaces. In particular, there are no pointwise derivatives $\xi^{(k)}$, except for $k < p - d/2$, since

$$\int\limits_{R^d} |\lambda|^{2k}(1+|\lambda|^{2p})^{-1}\, d\lambda \asymp \int\limits_{1}^{\infty} r^{2k+d-2p-1}\, dr = \infty$$

for $k \geq p - d/2$. And in a case of the dimension

$$d \geq 2p \tag{5.6}$$

the very ξ is essentially generalized random function. Moreover, in any bounded region $T \subseteq R^d$ its almost all realizations are no better than Schwartz distributions from the corresponding Sobolev space $\overset{\circ}{W}{}_2^{-q}(T)$, $q > d/2 - p$. To be more precise, for the Gaussian random function ξ we have it's realization in $\overset{\circ}{W}{}_2^{-q}(T)$ with probability 1 or 0, depending on $q > d/2 - p$ or $q \leq d/2 - p$.

Indeed, for $q > d/2 - p$, according to the known Hilbert–Schmidt embedding

$$\overset{\circ}{W}{}_2^{q}(T) \subseteq W_2^{-p}(T) \tag{5.6}$$

— see (4.27), there is a modification of ξ such that all its realizations are in the space $\overset{\circ}{W}{}_2^{-q}(T)$, dual to $\overset{\circ}{W}{}_2^{q}(T)$. With other hand, according to what we know about Gaussian random functions — see (2.2), if in a case

$$\mathrm{E}|(\varphi,\xi)|^2 \asymp \|\varphi\|_{-p}^2, \qquad \varphi \in C_0^\infty(T),$$

we have $\xi \in \overset{\circ}{W}{}_2^{-q}(T)$ with a positive probability, then we should have the Hilbert–Schmidt embedding (5.6), which holds true only for $q > d/2 - p$.

In spite of a lack of any visible "smoothness," any generalized random function $\xi \in \mathbf{W}_2^p(T)$ can be tested by all Schwartz distributions $x \in X(T) = W_2^{-p}(T)$. This possibility, in particular, let us learn certain things about the function ξ near boundary behaviour — we have in mind the certain limit properties related to the *generalized boundary trace*

$$\xi^{(k)} = (x, \xi^{(k)}) = (-1)^k (x \times \delta^{(k)}, \xi), \qquad x \in C_0^p(\Gamma), \quad k = 0, \dots, p-1, \tag{5.7}$$

on $\Gamma = \partial T$ of the generalized nontangent derivatives

$$\frac{\partial^k}{\partial l^k}\xi = \left(\varphi, \frac{\partial^k}{\partial l^k}\xi\right) = (-1)^k\left(\frac{\partial^k \varphi}{\partial l^k}, \xi\right), \qquad \varphi \in C_0^\infty(T \cap \Gamma^\varepsilon),$$

which are defined similar to what we know about the deterministic case for the regular boundary — cf. (4.13), etc. Here the generalized random functions

$$\xi^{(k)} = (x, \xi^{(k)}), \qquad x \in C_0^p(\Gamma),$$

on the boundary Γ can be characterized by their *meansquare* continuity with respect to the known norm $\|x\|_{-(p-k-1/2)}$, and with introduction of the corresponding *stochastic Sobolev spaces* $\mathbf{W}_2^{p-k-1/2}(\Gamma)$, this characterization means that

$$\xi^{(k)} \in \mathbf{W}_2^{p-k-1/2}(\Gamma), \qquad k = 0, \ldots, p-1,$$

with its norm

$$\|\xi^{(k)}\|_{p-k-1/2} = \sup_{\|x\|_{-(p-k-1/2)} \leq 1} (\mathrm{E}|(x, \xi^{(k)})|^2)^{1/2} \leq$$

$$\leq C \sup_{\|x \times \delta^{(k)}\|_{-p} \leq 1} (\mathrm{E}|(x \times \delta^{(k)}, \xi)|^2)^{1/2} \leq C\|\xi\|_p \tag{5.8}$$

— cf. (4.14). ■

In a similar way, for $\xi \in \mathbf{W}_2^{(p,1)}(T)$ from the anisotropic stochastic Sobolev space in the cylinder type region

$$T = G \times I \subseteq R^d$$

we can introduce the *generalized boundary trace*

$$\xi^{(k)} = (x, \xi^{(k)}) = (-1)^k (x \times \delta^{(k)}, \xi), \qquad x \in C_0^p(\partial G \times I),$$
$$k = 0, \ldots, p-1, \tag{5.9}$$

of the nontangent generalized derivatives on the side boundary $\partial G \times I$ and the *generalized boundary trace*

$$\xi^{(0)} = (x, \xi^{(0)}) = (x \times \delta^{(0)}, \xi), \qquad x \in C_0^\infty(G \times \partial I) \tag{5.9$'$}$$

of the very function ξ on the bottom/top boundary

$$G \times \partial I = (G \times \{a\}) \cup (G \times \{b\})$$

— cf. (4.23), (4.23)′.

Moreover, on the regular parts of the boundary this trace can be characterized as

$$\xi^{(k)} \in \mathbf{W}_2^{(p-k-1/2,1-k/p-1/(2p))}(\partial G \times I_{\text{loc}}), \qquad k = 0, \dots, p-1,$$

on the regular part of the side-boundary $\partial G \times I$ and

$$\xi^{(0)} \in \mathbf{W}_2^{p/2}(G_{\text{loc}})$$

on the regular parts $G_{\text{loc}} \times \{a\}$, $G_{\text{loc}} \times \{b\}$ of the bottom/top boundary $G \times \partial I$.

In particular, for

$$T = R^{d-1} \times (0, \infty),$$

say, we can characterize the boundary trace

$$\xi^{(0)} = (x, \xi^{(0)}) = (x \times \delta^{(0)}, \xi), \qquad x \in C_0^p(R^{d-1}),$$

in the whole as

$$\xi^{(0)} \in \mathbf{W}^{p/2}(R^{d-1}) \tag{5.10}$$

with its norm

$$\|\xi^{(0)}\|_{p/2} = \sup_{\|x\|_{-p/2} \leq 1} (\mathrm{E}|(x, \xi)|^2)^{1/2} \asymp$$

$$\asymp \sup_{\|x \times \delta^{(0)}\|_{-(p,1)} \leq 1} (\mathrm{E}|(x \times \delta^{(0)}, \xi)|^2)^{1/2} \leq C\|\xi\|_{(p,1)}.$$

Moreover, we can treat $\xi \in \mathbf{W}_2^{(p,1)}(T)$ as the vector-function

$$\xi = \xi_r, \qquad r \geq 0, \tag{5.11}$$

of the variable $r = t_2$ (with the variable $s = t_1 \in G$ hidden in ξ_r) where $\xi_r \in \mathbf{W}^{p/2}(R^{d-1})$ represents the generalized trace of ξ on the hyperplane $R^{d-1} \times \{r\} \subseteq R^{d-1} \times R^1$, determined as

$$\xi_r = (x, \xi_r) = (x \times \delta_r, \xi), \qquad x \in C_0^p(R^{d-1}),$$

with the delta-function δ_r at the point $r \in R^1$. And the vector-function (5.11) is such that

$$\|\xi_{r+h} - \xi_r\|_{p/2} \leq C\|x \times \delta_{r+h} - x \times \delta_r\|_{-(p,1)} \to 0, \qquad h \to 0,$$

with

$$\|x \times \delta_{r+h} - x \times \delta_r\|^2_{-(p,1)} \asymp$$

$$\asymp \iint\limits_{R^{d-1}\times R^1} |\tilde{x}(\lambda_1)|^2 |e^{-i\lambda_2 h} - 1|^2 (1 + |\lambda_1|^{2p} + \lambda_2^2)^{-1}\, d\lambda_1\, d\lambda_2$$

— see (5.4). This continuity of $\xi = \xi_r$, $r \geq 0$, let us apply the *stochastic integral representation*

$$\xi = (\varphi, \xi) = \int \varphi_r \xi_r \, dr, \qquad \varphi \in C_0^\infty(T), \tag{5.11$'$}$$

for $\xi \in \mathbf{W}_2^{(p,1)}(T)$ as the generalized random function in the region $T \subseteq R^d$. ■

5.3. Stochastic Dirichlet problem

Let

$$\mathcal{P} = \sum_{|k| \leq 2p} a_k \partial^k \geq 0$$

be a nondegenerate elliptic differential operator which gives the equivalent norm

$$\|\varphi\|_W = (\varphi, \mathcal{P}\bar{\varphi})^{1/2} \asymp \|\varphi\|_p, \qquad \varphi \in C_0^\infty(T_0),$$

in the Sobolev space $W = \overset{\circ}{W}{}_2^p(T_0)$ where $T_0 \supseteq [T]$ can be any region, containing the closure of T in R^d — cf. (3.18). Then, for any generalized random function $\xi \in \mathbf{W}_2^p(T)$ in the region T, we have

$$\eta = \mathcal{P}\xi \in \mathbf{W}_2^{-p}(T),$$

that is, the generalized random function

$$\eta = (\varphi, \eta) = (\mathcal{P}\varphi, \xi), \qquad \varphi \in C_0^\infty(T),$$

is meansquare continuous with respect to the corresponding norm

$$\|\varphi\|_p \asymp \|\mathcal{P}\varphi\|_{-p}$$

— see Section 4.2.

Let us consider the generalized stochastic differential equation

$$\mathcal{P}\xi = \eta \tag{5.12}$$

in the region T, having the regular boundary $\Gamma = \partial T$, with the *arbitrary* given $\eta \in \mathbf{W}_2^{-p}(T)$; the very equation means that

$$(\mathcal{P}\varphi, \xi) = (\varphi, \eta), \qquad \varphi \in C_0^\infty(T),$$

Let us set on the solution $\xi \in \mathbf{W}_2^p(T)$ we are looking for, the *Dirichlet type* stochastic *boundary conditions*

$$\xi^{(k)} = \xi_k, \qquad k = 0, \dots, p-1, \tag{5.13}$$

prescribing the nontangent derivatives $\xi^{(k)}$ of ξ on the boundary Γ, we know about, by means of *arbitrary* given generalized random functions

$$\xi_k \in \mathbf{W}_2^{p-k-1/2}(\Gamma), \qquad k = 0, \dots, p-1.$$

THEOREM. *There is the unique solution $\xi \in \mathbf{W}_2^p(T)$ of the stochastic Dirichlet problem* (5.12), (5.13).

PROOF. According to the direct sum representations (4.8), (4.15) we have

$$X(T) = W_2^{-p}(T) = \mathcal{P}\,\mathring{W}_2^p(T) \oplus \sum_{k=0}^{p-1} \oplus X^{(k)}(\Gamma)$$

with $\mathcal{P}\,\mathring{W}_2^p(T) = [\mathcal{P}C_0^\infty(T)]$ in $X(T)$. Hence, we have

$$x = \mathcal{P}u + \sum_{k=0}^{p-1} x_k \times \delta^{(k)}$$

$$(u \in \mathring{W}_2^p(T), \quad x_k \in W_2^{-(p-k-1/2)}(\Gamma), \quad k = 0, \dots, p-1)$$

$$\|x\|_{-p} \asymp \|u\|_p + \sum_{k=0}^{p-1} \|x_k\|_{-(p-k-1/2)}$$

for all $x \in X(T) = W_2^{-p}(T)$. The generalized random function $\eta = (\varphi, \eta)$, $\varphi \in C_0^\infty(T)$, is well-defined on the Sobolev space $\mathring{W}_2^p(T) = [C_0^\infty(T)]$,

$$(u, \eta) = \lim_{\varphi \to u} (\varphi, \eta), \qquad u \in \mathring{W}_2^p(T),$$

and we can set

$$\xi = (x, \xi) \overset{\text{def}}{=} (u, \eta) + \sum_{k=0}^{p-1} (x_k, \xi_k), \qquad x \in X(T) = [C_0^\infty(T)]. \tag{5.14}$$

Obviously, it is a generalized random function on $X(T) = W_2^{-p}(T)$, that is, $\xi \in \mathbf{W}_2^p(T)$. To explain, for the meansquare norm of all random variables in (5.14) we have

$$\|(x,\xi)\| \le \|(u,\eta)\| + \sum_{k=0}^{p-1} \|(x_k,\xi_k)\| \le$$

$$\le C\Big(\|u\|_p + \sum_{k=0}^{p-1} \|x_k\|_{-(p-k-1/2)}\Big) \le C\|x\|_{-p}, \qquad x \in X(T).$$

The formula (5.14), in particular, gives

$$(\mathcal{P}\varphi, \xi) = (\varphi, \eta), \qquad \varphi \in C_0^\infty(T),$$

and

$$(x \times \delta^{(k)}, \xi) = (x_k, \xi^{(k)}) = (x_k, \xi_k),$$
$$x \in \mathring{W}_2^{-(p-k-1/2)}(\Gamma), \quad k = 0, \dots, p-1,$$

so it gives the solution ξ. Obviously, any solution $\xi \in \mathbf{W}_2^p(T)$ can be described by means of the very formula (5.14).

Actually, this formula describes the certain structure of all generalized random functions from the stochastic Sobolev space $\mathbf{W}_2^p(T)$. ■

In a similar way we can describe anisotropic stochastic Sobolev spaces. We illustrate it with the stochastic Sobolev space $\mathbf{W}_2^{(p,1)}(T)$ in the region

$$T = R^{d-1} \times (0, \infty).$$

Having in mind the known structure of the test functions space $X(T) = W_2^{-(p,1)}(T)$ — see Section 4.4, let us take any nondegenerate elliptic operator $Q = \sum_{|k| \le 2p} a_k \partial^k$, related to the variable $t_1 = s \in R^{d-1}$, and set

$$\mathcal{P} = -\frac{\partial^2}{\partial r^2} + Q \ge 0$$

with $r = t_2 \in R^1$ — cf. (3.19).

For all $\xi \in \mathbf{W}_2^{(p,1)}(T)$ we have

$$\eta = \mathcal{P}\xi \in \mathbf{W}^{-(p,1)}(T),$$

that is, the generalized random function

$$\eta = (\varphi, \eta) = (\mathcal{P}\varphi, \xi), \qquad \varphi \in C_0^\infty(T),$$

is meansquare continuous with respect to the corresponding norm

$$\|\varphi\|_{(p,1)} \asymp \|\mathcal{P}\varphi\|_{-(p,1)}.$$

Let us consider the generalized stochastic differential equation

$$-\frac{\partial^2}{\partial r^2}\xi + Q\xi = \eta \tag{5.15}$$

in the cylinder $T = R^{d-1} \times (0, \infty)$ with the arbitrary given $\eta \in \mathbf{W}_2^{-(p,1)}(T)$, and set the *stochastic boundary condition*

$$\xi^{(0)} = \xi_0 \tag{5.16}$$

on the generalized solution $\xi \in \mathbf{W}_2^{(p,1)}(T)$, prescribing the boundary trace of ξ on

$$\Gamma = R^{d-1} \times \{0\} \quad (= R^{d-1})$$

at the initial $r = 0$ by means of the arbitrary given

$$\xi_0 \in \mathbf{W}^{p/2}(R^{d-1})$$

— cf. (5.10), (5.11).

THEOREM. *There is the unique solution* $\xi \in \mathbf{W}_2^{(p,1)}(T)$ *of the stochastic boundary problem* (5.15), (5.16).

Similar to (5.14), this solution can be described as

$$\xi = (x, \xi) \overset{\text{def}}{=} (u, \eta) + (x_0, \xi_0),$$

$$x = \mathcal{P}u + (x_0 \times \delta^0) \in X(T) = [C_0^\infty(T)],$$

with the corresponding

$$u \in \overset{\circ}{W}{}_2^{(p,1)}(T) = [C_0^\infty(T)], \qquad x_0 \in W_2^{-p/2}(R^{d-1}),$$

thanks to the direct sum representation

$$X(T) = W_2^{-(p,1)}(T) = \mathcal{P}\, \overset{\circ}{W}{}_2^{(p,1)}(T) + X(\Gamma)$$

with the known boundary tests functions space $X(\Gamma)$.

Chapter II

Differential Equations for Generalized Random Functions

II.1. Generalized Differential Equations

1.1. Test functions for generalized differential equations

A linear differential operator $\mathcal{L} = \sum a_k \partial^k$ (with infinitely differentiable coefficients, say) acts on the generalized functions $u = (\varphi, u)$, $\varphi \in D = C_0^\infty(T)$, in the region $T \subseteq R^d$ as

$$\mathcal{L}u = (\varphi, \mathcal{L}u) \equiv (\mathcal{L}^*\varphi, u), \qquad \varphi \in D,$$

where

$$\mathcal{L}^*\varphi = \sum (-1)^{|k|} \partial^k (a_k \varphi).$$

Of course, for sufficiently regular functions the differential operator $\mathcal{L}$ can act directly as

$$\mathcal{L}u = \sum_k a_k \partial^k u.$$

Let us consider the equation

$$\mathcal{L}u = f, \tag{1.1}$$

where on the right-hand side generalized function $f = (\varphi, f)$, $\varphi \in D$, is characterized by continuity over the test functions $\varphi \in D$ with respect to the norm $\|\varphi\|_F$ in some Hilbert space F (for example, $F = L_2(T)$). Having in mind that f can be any representative from a class of all functions of this kind, let us choose an appropriate functional class **W** of generalized functions $u = (\phi, u)$, $\phi \in D$, associated with equation (1.1).

Equation (1.1) in the region $T \subset R^d$ means that

$$(\mathcal{L}^*\varphi, u) = (\varphi, f) \tag{1.1$'$}$$

for all test functions $\varphi \in D = C_0^\infty(T)$. According to (1.1)$'$ with $f = (\varphi, f)$, $\varphi \in D$, continuous with respect to $\|\varphi\|_F$, we get $u = (x, u)$, $x = \mathcal{L}^*\varphi$, continuous with respect to the corresponding $\|x\| = \|\varphi\|_F$.

The problem is to find an appropriate extension of $u = (x, u)$, $x \in \mathcal{L}^*D \subseteq D$, up to a generalized function $u = (x, u)$, $x \in D = C_0^\infty(T)$.

Let us apply a closure $[D]$ of $D \subseteq F$ in F, assuming

$$[D] = F \subseteq D^*,$$

where the imbedding $F \subseteq D^*$ is such that (strong) convergence $g_n \to g$ in the Hilbert space F implies (weak) convergence $g_n \to g$ in the Schwartz distribution space D^*, i.e.,

$$(\varphi, g_n) \to (\varphi, g), \qquad \varphi \in D = C_0^\infty(T).$$

Along with the convergence $g_n \to g$ in D^* we also have $\mathcal{L}g_n \to \mathcal{L}^*g$ according to

$$(\varphi, \mathcal{L}^*g_n) = (\mathcal{L}\varphi, g_n) \to (\mathcal{L}\varphi, g) = (\varphi, \mathcal{L}^*g), \qquad \varphi \in D.$$

In equations (1.1)–(1.1)$'$ we can treat $f = (g, f)$ with $g \in F$ regarded as test functions; here we have in mind a continuous extension of the generalized function $f = (g, f)$, $g \in D$, to the wider test functions space $[D] = F$ (it is continuous with respect to $\|g\|_F$, $g \in F$). For any $g \in F$ regarded as the limit $g = \lim \varphi$ of $\varphi \in D$, we have the corresponding limit

$$x = \mathcal{L}^*g = \lim(\mathcal{L}^*\varphi \in D^*)$$

in equation (1.1)$'$ such that

$$(x, u) = \lim(\mathcal{L}^*\varphi, u) = \lim(\varphi, \mathcal{L}u) = (g, f),$$

where (x, u) is determined by the right-hand side of this equation and it can be treated as the result of *testing* $u \in \mathbf{W}$ by the *generalized test functions* $x = \mathcal{L}^*g$, $g \in F$. According to this interpretation of

$(x, u) = (g, f)$ we obtain the same result for various $g \in F$ with the same $x = \mathcal{L}^* g = \lim(\mathcal{L}^* \varphi)$ in D^*; it is worth mentioning that while dealing with these $x \in D^*$ as generalized functions

$$x = (\varphi, x), \qquad \varphi \in D = C_0^\infty(T),$$

we do not make any distinction between functions that coincide in the considered region $T \subseteq R^d$ and, in particular, we have

$$(x, u) = 0, \qquad x = \mathcal{L}^* g$$

for all $g \in F$ such that

$$(\varphi, \mathcal{L}^* g) = (\mathcal{L}\varphi, g) = 0, \qquad \varphi \in D.$$

As a result of this approach to equation (1.1), we get a linear function $u = (x, u)$, $x \in \mathcal{L}^* F \subseteq D^*$, as

$$(\mathcal{L}^* g, u) = (g, f), \qquad g \in F. \tag{1.2}$$

This requires f to be such that

$$(g, f) = 0$$

with $\mathcal{L}^* g = 0$ in D^*. Obviously all $g \in F$ with $\mathcal{L}^* g = 0$ in D^* form a (closed) subspace in our Hilbert space F. Let $\overset{\circ}{F}$ be an orthogonal complement to this subspace. Then, for any $g \in F$ and its orthogonal projection $\overset{\circ}{g}$ onto $\overset{\circ}{F}$ we have $g - \overset{\circ}{g}$ in the orthogonal complement to $\overset{\circ}{F}$, i.e., $\mathcal{L}^*(g - \overset{\circ}{g}) = 0$; hence $(g - \overset{\circ}{g}, f) = 0$ and

$$f = (g, f) = (\overset{\circ}{g}, f), \qquad g \in F,$$

must be continuous with respect to $\| \overset{\circ}{g} \|_F \leq \|g\|_F$. We call any f of this kind as *admissible function* for the equation (1.1). In a case of the *nondegenerate* operator $\mathcal{L}^* \colon F \to D^*$ we have $\overset{\circ}{F} = F$ ($\mathcal{L}^* g = 0$ if and only if $g = 0$). In a general case, the operator

$$\mathcal{L}^* \colon\ F \to D^*$$

transfers the space Hilbert F to the space

$$X = \mathcal{L}^* F = \mathcal{L}^* \ \overset{\circ}{F} \subset D^*, \tag{1.3}$$

and from the equations (1.1)–(1.2) we find that $u = (x, u)$ is continuous on the generalized test functions $x \in X$ with respect to a norm

$$\|x\|_X = \| \overset{\circ}{g} \|_F, \qquad x = \mathcal{L}^* \overset{\circ}{g}, \qquad \overset{\circ}{g} \in \overset{\circ}{F}.$$

Obviously, we have the Hilbert space

$$X = \mathcal{L}^* F = \mathcal{L}^*[D] = [\mathcal{L}^* D]$$

as a closure of $\mathcal{L}^* D$, with $\mathcal{L}^*$ on $\overset{\circ}{F}$ as a *unitary* operator; of course in the case of the *nondegenerate* operator $\mathcal{L}^*$ we have

$$\|x\|_X = \|g\|_F, \qquad x = \mathcal{L}^* g, \qquad g \in F.$$

In our scheme, the bilinear form (φ, g), $\varphi \in D$, $g \in F$, is continuous with respect to $\|g\|_F$; therefore

$$\sup_{\|g\|_F \leq 1} |(\varphi, g)| < \infty$$

and, in particular, any $\varphi \in D$ determines an element $\varphi \in W$ of a dual space $W = X^*$

$$(\varphi, x) = (\mathcal{L}\varphi, \overset{\circ}{g}), \qquad x = \mathcal{L}^* \overset{\circ}{g} \in X,$$

with the norm

$$\|\varphi\|_W = \sup_{\|\overset{\circ}{g}\|_F \leq 1} |(\mathcal{L}\varphi, \overset{\circ}{g})|. \tag{1.4}$$

Moreover, here $W = [D]$ is a closure of $D \subseteq W$. Indeed, $(\varphi, x) = 0$, $\varphi \in D$, implies $x = 0$ for any generalized function $x \in X = W^* \subset D^*$.

Using $W = [D]$, we have

$$X = W^*,$$

i.e., $x \in X$ can be characterized as the generalized functions $x = (\varphi, x)$, $\varphi \in D$, which are continuous with respect to $\|\varphi\|_W$.

Let $W = [D]$ be such that

$$W \subseteq D^* \tag{1.5}$$

and (strong) convergence $\varphi \to 0$ in W implies (weak) convergence $\varphi \to 0$ in the Schwartz distributions space D^*, i.e., $(x, \varphi) \to 0$ for any $x \in D$. Then, for all $x \in D$ we have $(x, \varphi) \equiv (\varphi, x)$, $\varphi \in D$, continuous with respect to the norm $\|\varphi\|_W$. Hence, $D \subseteq X$ and with $\mathcal{L}^*D \subseteq D$, $[\mathcal{L}^*D] = X$ we have

$$X = \mathcal{L}^*F = [D] \tag{1.6}$$

being the closure of $D = C_0^\infty(T)$. And in a case when the norm

$$\|x\|_X = \sup_{\|\varphi\|_W \leq 1} (\varphi, x), \qquad \varphi \in D,$$

is continuous over $x \in D$ in the Schwartz space $D = C_0^\infty(T)$, the limit formula (1.2) which determines $u = (x, u)$ for all test functions $x \in X$ defines the corresponding generalized function

$$u = (x, u), \qquad x \in D = C_0^\infty(T) \subseteq X$$

(To illustrate this case one can assume that $W \subseteq L_2(T)$, say; for example, it implies that $X = W^* \supseteq L_2(T)$ with $\|x\|_X \leq C\|x\|_{L_2}$, $x \in D = C_0^\infty(T)$.)

It seems natural to associate with equation (1.1) a functional class **W** characterized by the continuity of its representatives $u = (\varphi, u)$, $\varphi \in D$, with respect to the norm $\|\varphi\|_X$ in $X \subset D^*$.

The assumptions made actually mean that we have the known W-type spaces

$$W = [D] = X^* \subseteq D^*, \qquad X = [D] = W^* \subseteq D^*,$$

and if it takes place we are getting the following result.

THEOREM. *The unique solution $u \in W$ of the equation* (1.1) *does exist for any admissible function in its right-hand side. This solution can be described by the limit diagram*

$$\begin{array}{ccccc} (\mathcal{L}^*\varphi, u) & = & (\varphi, f), & & \varphi \in D, \\ \downarrow & & \downarrow & & \\ (\mathcal{L}^*g, u) & = & (g, f), & & g \in F, \end{array}$$

with $x = \mathcal{L}^*g$ running over the test functions space $X = \mathcal{L}^*F[D]$. ∎

In what follows, we deal with generalized *random* functions which can be *deterministic* as well; to stress on randomness we use notations like $u = \xi$, $f = \eta$, etc. Speaking on continuity of *random* functions we have in mind their meansquare continuity. It means, in particular, that the functional class **W** associated with the equation (1.1) consists of all generalized random functions $u = \xi$, having meansquare values $\|(\varphi,\xi)\| = (\mathrm{E}|(\varphi,\xi)|^2)^{1/2}$ continuous with respect to the norm $\|\varphi\|_X$ over $\varphi \in D = C_0^\infty(T)$. And the solution $u = \xi \in \mathbf{W}$ can be determined by the meansquare limit in the diagram suggested.

The following remark seems to be important.

As we know, any generalized random function can be treated as a random Schwartz distribution with all realizations in D^*. And treating in this way the solution $u = \xi \in \mathbf{W}$ of the equation (1.1) and the generalized random function $f = \eta$, we with the probability 1 have

$$(\mathcal{L}^*\varphi,\xi) = (\varphi,\eta)$$

for a countable complete system $\{\varphi\}$ in the Schwartz space $D = C_0^\infty(T)$, i.e.,

$$(\mathcal{L}^*\varphi,\xi_\omega) = (\varphi,\eta_\omega)$$

for almost all realizations ξ_ω, η_ω being Schwartz distributions. These ξ_ω, η_ω are continuous over $\varphi \in D$ in the Schwartz space $D = C_0^\infty(T)$ and the equation, we have for the complete system $\{\varphi\}$ in D, can be extended over all $\varphi \in D$ giving us the generalized differential equation (1.1) for $u = \xi_\omega$, $f = \eta_\omega \in D^*$. Any other realizations ξ_ω, η_ω appear only with the probability 0, and we can take the new modifications of ξ,η with the redefined $\xi_\omega = 0$, $\eta_\omega = 0$, say. Then *we have the equation* (1.1) *for all realizations* $u = \xi_\omega$, $f = \eta_\omega$.

Of course, if this holds true, then we have the equation (1.1) for $u = \xi$, $f = \eta$ as generalized random functions. Thus, we can equivalently consider equation (1.1) for the realizations $u = \xi_\omega$, $f = \eta_\omega$. But in the latter case the diagram, suggested for equation (1.1) solution, cannot be applied because we have no the required continuity of $f = \eta_\omega$, $u = \xi_\omega$ over the test functions space D. ■

EXAMPLE. (General operators in the space $F = L_2(T)$). Let us consider equation (1.1), (1.2) with the general differential operator $\mathcal{L}$ and the right-hand side which corresponds to the Hilbert space $F = L_2(T)$. In this case it could be nontrivial $g \in F$, $\mathcal{L}^*g = 0$, which form alltogether an orthogonal complement to a closure

$$\overset{\circ}{F} = [\mathcal{L}D]$$

in $F = L_2(T)$, since $\mathcal{L}^* g = 0$ in D^* means that

$$\overline{(\varphi, \mathcal{L}^* g)} = \langle \mathcal{L}\bar{\varphi}, g \rangle_{L_2} = 0, \qquad \varphi \in D = C_0^\infty(T).$$

Obviously, in (1.4) we have the norm

$$\|\varphi\|_W = \|\mathcal{L}\varphi\|_{L_2} = (\varphi, \mathcal{P}\bar{\varphi})^{1/2}, \qquad \varphi \in D,$$

with the symmetric positive operator

$$\mathcal{P} = \mathcal{L}^* \mathcal{L},$$

and the corresponding space $W = [D]$ appears as in our scheme of Section 3.1, Chapter I.

Hence, the generalized test functions space $X = W^*$ for the equation (1.1), (1.2) is

$$X = \mathcal{L}^* L_2(T) = \mathcal{P} W. \tag{1.7}$$

Considering $\mathcal{P} = \mathcal{L}^* \mathcal{L}$, one can apply $\mathcal{L}$ as well defined on the whole $W = [D]$, being a unitary extension

$$\mathcal{L}\colon\ W = [D] \to \overset{\circ}{F} = [\mathcal{L}D]$$

from $D = C_0^\infty(T)$. And the property of $\mathcal{L}^*$ to be nondegenerate (unitary) operator

$$\mathcal{L}^*\colon\ F = L_2(T) \to X$$

can be characterized as

$$[\mathcal{L}D] = L_2(T). \tag{1.8}$$

One can observe here (with a little bit of a surprise, maybe) that the differential equation (1.1) for $u \in \mathbf{W}$ is equivalent to the equation

$$\mathcal{P}u = \mathcal{L}^* f$$

which appears after applying the differential operator $\mathcal{L}^*$ to (1.1). It can be seen in our framework from the corresponding diagram

$$\begin{array}{ccll} (\mathcal{L}^* \overset{\circ}{g}, u) & = & (\overset{\circ}{g}, f), & \overset{\circ}{g} = \mathcal{L}v \in \overset{\circ}{\mathbf{F}} = [\mathcal{L}D], \\ \| & & & \\ (\mathcal{P}v, u) & = & (v, \mathcal{L}^* f), & v \in W. \quad \blacksquare \end{array}$$

EXAMPLE. (Symmetric positive operators in the Hilbert space F of W-type). Let us consider equation (1.1), (1.2) with the symmetric positive differential operator

$$\mathcal{L} = \mathcal{L}^* = \mathcal{P} \geq 0 \tag{1.6'}$$

and the right-hand side which corresponds to the Hilbert space

$$F = W = [D]$$

associated with the operator $\mathcal{P}$ according to its norm

$$\|\varphi\|_W = (\varphi, \mathcal{P}\varphi)^{1/2}, \qquad \varphi \in D.$$

Obviously, this is the norm appeared in (1.4) with $\overset{\circ}{g} = g$,

$$(\mathcal{L}\varphi, g) = (\varphi, \mathcal{P}g) = \langle \varphi, \bar{g} \rangle_W$$

and the generalized test functions space (1.3) for equations (1.1), (1.2) is

$$X = \mathcal{L}^* F = \mathcal{P}W \tag{1.7'}$$

which exactly fits our scheme of Section 3.1, Chapter 1 with

$$\mathcal{L}^* = P\colon\ F = W \to X \subset D^* \tag{1.9}$$

as a nondegenerate (*unitary*) operator. Recall, there were given some conditions for the space X to be of the required W-type (see (3.2), (3.5)). ■

In what follows, we are dealing with suggested above scheme for the general operators $\mathcal{L}$ in $F = L_2(T)$ or the symmetric positive operators $\mathcal{L} = \mathcal{L}^* = \mathcal{P}$.

1.2. Generalized differential equations for some probability functional models

Let us consider in the framework of our scheme with the space $F = L_2(T)$ the ordinary differential operator $\mathcal{L} = d/dt$ and the corresponding equation

$$\frac{d}{dt}u = f \tag{1.10}$$

on a finite or infinite interval $T \subseteq R^1$ with the initial $t = 0$, say. Note that any generalized function $u\colon \mathcal{L}u = f$ for $\mathcal{L} = d/dt$ can be described as

$$(x, u) = (x, C) + (\varphi, f), \qquad x \in D,$$

with a constant $C = (x_0, u)$ and $\varphi(t) = -\int_a^t y(s)\,ds$ from $D = C_0^\infty(T)$, $\mathcal{L}^*\varphi = y = x - (x,1)x_0$ for $\mathcal{L}^* = -d/dt$, where x_0 is just some $x_0 \in D$, $(x_0, 1) = 1$.

Taking into account that any $x \in D = C_0^\infty(T)$ is represented as

$$x = \mathcal{L}^* g = -\frac{d}{dt} g,$$

$$g(t) = \int_T x(s)\,ds - \int_a^t x(s)\,ds, \qquad t \in T,$$

with $g \in L_2(T)$, we immediately see that our generalized test functions space $X = \mathcal{L}^* L_2(T)$ is such that

$$D \subseteq X, \qquad X = [D].$$

In a case of the infinite interval $T = (0, \infty)$ we have a nondegenerate operator $\mathcal{L}^*: L_2(T) \to X$, since a generalized solution of $\mathcal{L}^* g = -dg/dt = 0$ is $g = \text{const}$ and the only $g \in L_2(T)$ is $g = 0$. The complete system of delta-functions $x = \delta_t = \mathcal{L}^* g_t$ in X labelled by $t \in T$, arises from the corresponding system of indicator functions $g_t = 1_{(0,t)} \in L_2(T)$, and any $u \in \mathbf{W}$ can be described as

$$u = u(t) \equiv (\delta_t, u) = (1_{(0,t)}, f), \qquad t \in T,$$

with

$$\lim_{t \to 0} \delta_t = 0, \qquad \lim_{t \to 0} u(t) = u(0) = 0.$$

Here, in the case of the equation (1.10) with a stochastic source f represented by the Gaussian white noise on the $L_2(T)$-space, we have the unique solution $u = \xi \in \mathbf{W}$ representing the well-known *Brownian motion* $\xi = \xi(t)$, $t \geq 0$, on the interval $T = (0, \infty)$.

Note, in a case of the finite interval $T = (0, 1)$, say, with $\mathcal{L}^* g = -dg/dt = 0$ for $g = 1 \in L_2(T)$, the corresponding subspace $\overset{\circ}{F} \subseteq F = L_2(T)$ consists of all $\overset{\circ}{g} \in L_2(T)$, $\int \overset{\circ}{g}(t)\,dt = 0$.

As was mentioned in general, equation (1.10) for $u \in \mathbf{W}$ is equivalent to the equation

$$-\frac{d^2}{dt^2} u = f \tag{1.11}$$

on the interval T with the symmetric positive operator $-\frac{d^2}{dt^2} = (\frac{d}{dt})^*(\frac{d}{dt})$ and the new right-hand side which appears as a result of applying the operator $\mathcal{L}^* = -d/dt$ to (1.10). One can explain this simply by employing such properties of $u \in \mathbf{W}$ as

$$\lim_{t\to 1} u(t) = 0$$

with $\lim_{t\to 1} \delta_t = 0$ in the case of $T = (0,1)$, say, and

$$\mathrm{E}|u(t)|^2 = \mathrm{E}|(1_{(0,t)}, f)|^2 \le C\|1_{(0,t)}\|^2_{L_2(T)} = Ct, \qquad t \to \infty,$$

in the case $T = (0, \infty)$.

For equation (1.10) with the Gaussian white noise on the L_2-space we get in (1.11) the new stochastic source $f = \eta$ being a Gaussian white noise on the corresponding W-space

$$\mathrm{E}(\varphi, \eta) = 0, \quad \mathrm{E}|(\varphi, \eta)|^2 = \|\varphi\|^2_W = \left\|\frac{d}{dt}\varphi\right\|^2_{L_2}, \quad \varphi \in W = [D],$$

and the Brownian motion $\xi = \xi(t)$, $t \ge 0$, is the unique solution $u = \xi \in \mathbf{W}$ of equation (1.11) on $T = (0, \infty)$ with this white noise on the W-space. The same equation (1.11) on the finite interval $T = (0,1)$ brings us the well-known *Brownian bridge* $\xi = \xi(t)$, $0 \le t \le 1$, as the unique solution $u = \xi \in \mathbf{W}$; recall that it arises as Brownian motion conditioned with

$$\xi(0) = \xi(1) = 0$$

at the boundary points of $T = (0,1)$, and can be described by means of the very Brownian motion $\xi_0(t)$, $t \ge 0$, with $\xi_0(0) = 0$ as

$$\xi(t) = \xi_0(t) - t\xi_0(1), \qquad 0 \le t \le 1. \quad \blacksquare$$

Let us consider the equation

$$\frac{\partial^2}{\partial t_1 \partial t_2} u = f \tag{1.12}$$

with $t = (t_1, t_2)$ in the plain region $T = (0, \infty) \times (0, \infty)$ and apply our general approach to the operator $\mathcal{L} = \frac{\partial^2}{\partial t_1 \partial t_2}$ in the space $F = L_2(T)$. The operator $\mathcal{L}^* = \frac{\partial^2}{\partial t_1 \partial t_2}$ is nondegenerate since for a generalized solution $g \in L_2(T)$ of $\mathcal{L}^* g = 0$ in T we have

$$(\mathcal{L}\varphi, g) = \int_0^\infty \int_0^\infty \varphi_1'(t_1)\varphi_2'(t_2) g(t)\, dt_1\, dt_2 = 0$$

for arbitrary $\varphi = \varphi_1 \times \varphi_2 \in C_0^\infty(T)$ which implies

$$g_1(t_1) = \int_0^\infty \varphi_2'(t_2) g(t)\, dt_2 \equiv 0$$

for a generalized solution $g_1 \in L_2(T_1)$ of $dg_1/dt_1 = 0$ on the infinite interval $T_1 = (0, \infty)$, and therefore $g(t_1, t_2) \equiv 0$ is the unique generalized solution in $L_2(T_2)$ of $dg/dt_2 = 0$ on the infinite interval $T_2 = (0, \infty)$. Hence our test functions space $X = \mathcal{L}^* L_2(T)$ for the equation (1.12) consists of all generalized functions

$$x = \mathcal{L}^* g = \frac{\partial^2}{\partial t_1 \partial t_2} g \in D^*, g \in L_2(T),$$

with the norm $\|x\|_X = \|g\|_{L_2}$. Obviously, operator $\mathcal{P} = \mathcal{L}^* \mathcal{L}$ satisfies the condition (3.2) of Chapter I, and for the corresponding W-space we have $X = W^* = [D]$, $D = C_0^\infty(T)$. Along with $x \in D$ our test functions space $X = \mathcal{L}^* L_2(T)$ contains delta-functions $x = \delta_t$,

$$\delta_t = \mathcal{L}^* 1_{(0,t_1) \times (0,t_2)}$$

with indicator functions of $(0, t_1) \times (0, t_2) \subseteq T$. These $x = \delta_t$, labelled by $t \in T$, form a complete system in X since the corresponding indicator functions form a complete system in $L_2(T)$. Therefore, any generalized random function $u \in \mathbf{W}$ can be described in terms of $f = \mathcal{L}u$ as

$$u = u(t) \equiv (\delta_t, u) = (1_{(0,t_1) \times (0,t_2)}, f), \qquad t = (t_1, t_2) \in T.$$

All $f = \mathcal{L}u$, $u \in \mathbf{W}$, can be characterized as generalized random functions $f = (\varphi, f)$, $\varphi \in D = C_0^\infty(T)$, which are meansquare continuous with respect to $\|\varphi\|_{L_2}$ and well-defined as $f = (g, f)$, $g \in L_2(T)$. In the deterministic case we have $f = \mathcal{L}u$ for $u \in W = X^*$ as all $f \in L_2(T)$ where according to our general scheme $W = [D]$ appears with the norm $\|u\|_W = \|f\|_{L_2}$. All $u \in \mathbf{W}$ satisfy the zero boundary conditions

$$\lim_{t_1 \to 0} u(t_1, t_2) = u(0, t_2) = 0, \qquad \lim_{t_2 \to 0} u(t_1, t_2) = u(0, t_1) = 0.$$

Equation (1.12) with the Gaussian white noise on the $L_2(T)$-space in the right-hand side describes, as its unique solution $u = \xi \in$

W, the well known *Brownian Sheet*

$$\xi(t) = \int_0^{t_1}\int_0^{t_2} \eta(ds), \qquad t \in t,$$

associated with the corresponding stochastic measure with independent values,

$$E\eta(dt) = 0, \qquad E|\eta(dt)|^2 = dt.$$

It is worth mentioning that when the space variable $s = 1/\sqrt{2}(t_1 - t_2)$ and the time variable $r = 1/\sqrt{2}(t_1 + t_2)$ are introduced, equation (1.12) represents the well-known *String equation* (known also as the *Wave equation*)

$$\frac{\partial^2 u}{\partial r^2} - \frac{\partial^2 u}{\partial s^2} = f \qquad (1.12)'$$

in a new region $\{(s,r)\colon |s| < r,\, r > 0\}$ bounded by the characteristics $|s| = r$, $r > 0$, of this equation, where f is of the same type as in the original equation (1.12). Our zero boundary conditions for $u \in \mathbf{W}$ describe the well-known *Goursat boundary problem* for the String equation. ■

Let us consider equation (1.1) with the Laplace operator $\mathcal{L} = \Delta$,

$$\Delta u = f \qquad (1.13)$$

in the region $T = R^3 \backslash \{0\}$ of all $t \in R^3$ but $t = 0$,

$$\Delta = \sum_k \frac{\partial^2}{\partial t_k^2}.$$

In the framework of our approach to $\mathcal{L} = \Delta$ in the $L_2(T)$-space we get the test functions space $X = \mathcal{L}^* L_2(T)$ for equation (1.13) as a result of the nondegenerate mapping $\mathcal{L}^* = \Delta\colon L_2(T) \to X$. Indeed, the generalized function $x = \mathcal{L}^* g$ with $\mathcal{L}^* g = 0$ in $T = R^3 \backslash 0$, $g \in L_2(R^3)$, could have its support $\operatorname{supp} x$ only at the point $t = 0$, its Fourier transform could only be a polynomial $\tilde{x}(\lambda) = |\lambda^2| \tilde{g}(\lambda)$ and, because of the correspondingly required behaviour of $\tilde{g}(\lambda) = |\lambda|^{-2}\tilde{x}(\lambda)$ at $\lambda \to 0$ and $\lambda \to \infty$ we have $\tilde{g} \in L_2(R^3)$ only in the case of $\tilde{g} = 0$. Thus, in our scheme we can consider equation (1.13) with any generalized random function $f = (\varphi, f)$, $\varphi \in D = C_0^\infty(T)$, meansquare continuous with respect to $\|\varphi\|_{L_2}$ in the $L_2(T)$-space.

The verified triviality of $g: \mathcal{L}^* g = 0$ in $L_2(T)$ also means that the collection $\mathcal{L}C_0^\infty(T)$ of all functions $\mathcal{L}u = \Delta u$, $u \in C_0^\infty(T)$, is dense in $L_2(T) = L_2(R^3)$. Therefore, the Fourier transforms

$$\alpha(\lambda) = \widetilde{\Delta u} = |\lambda|^2 \tilde{u}(\lambda), \qquad \lambda \in R^3,$$

are dense in the dual $L_2(R^3)$. Hence, in the representation

$$u(t) = \frac{1}{(2\pi)^{3/2}} \int\limits_{R^3} \frac{e^{i\lambda t} - 1}{|\lambda|^2} \alpha(\lambda)\, d\lambda, \qquad t \in R^3, \tag{1.14}$$

which appears as the inverse Fourier transform

$$u(t) = \frac{1}{(2\pi)^{3/2}} \int\limits_{R^3} e^{i\lambda t} \tilde{u}(\lambda)\, d\lambda$$

for $u \in C_0^\infty(T)$, $T = R^3 \backslash 0$, with

$$u(0) = \frac{1}{(2\pi)^{3/2}} \int\limits_{R^3} \frac{1}{\lambda^2} \alpha(\lambda)\, d\lambda = 0,$$

we can go to any limit $\alpha \in L_2(R^3)$, and this leads to the representation (1.14) of all $u \in W$ from the dual space $W = [C_0^\infty(T)] = X^*$ with the norm

$$\|u\|_W = \|\Delta u\|_{L_2} = \|\alpha\|_{L_2}.$$

One can apply a system of functions $g_t \in L_2(T)$ labelled by $t \in R^3$, which appears thanks to the Fourier transform of

$$\tilde{g}_t(\lambda) = \frac{1}{(2\pi)^{3/2}} \frac{e^{i\lambda t} - 1}{|\lambda|^2}, \qquad \lambda \in R^3,$$

in the representation (1.14). Namely, the corresponding system $\tilde{g}_t \in L_2(R^3)$ is complete, since for $\alpha \in L_2(R^3)$

$$\int [\tilde{g}_{t+s}(\lambda) - \tilde{g}_t(\lambda)] \alpha(\lambda)\, d\lambda = \frac{1}{(2\pi)^{3/2}} \int e^{i\lambda t} \Big[\frac{e^{i\lambda s} - 1}{|\lambda|^2} \alpha(\lambda) \Big]\, d\lambda \equiv 0$$

only if $\alpha = 0$. According to (1.14) for $u = \varphi \in D$, we have

$$\varphi(t) = (\tilde{g}_t, |\lambda|^2 \tilde{\varphi}) = (g_t, \Delta\varphi) = (\varphi, \mathcal{L}^* g_t)$$

which shows that in our test functions space $X = \mathcal{L}^* L_2(T)$ we have a complete system of delta-functions $x = \delta_t = \mathcal{L}^* g_t$, $t \in T$, describing all $u \in \mathbf{W}$

$$u = u(t) = (\delta_t, u) = (g_t, f), \qquad t \in T, \tag{1.15}$$

in terms of $f = \mathcal{L}u$.

According to (1.15), the random function $u = u(t)$, $t \in T$, is meansquare continuous and satisfies the boundary condition

$$\lim_{t \to 0} u(t) = u(0) = 0$$

at the boundary point $t = 0$ of $T = R^3 \backslash \{0\}$. In the case of the Gaussian white noise $f = \eta$ in the $L_2(T)$-space, the unique solution $u = \xi \in \mathbf{W}$ represents the known *Levy Brownian motion*,

$$\mathrm{E}\xi(t) = 0, \qquad \mathrm{E}|\xi(t)|^2 = \sigma^2 |t|, \qquad t \in R^3,$$

with the corresponding diffusion coefficient σ^2. To verify it one can apply an orthogonal transformation $\lambda \to \mu$ with the component

$$\mu_1 = \lambda_1 \frac{t_1}{|t|} + \lambda_2 \frac{t_2}{|t|} + \lambda_2 \frac{t_3}{|t|}$$

to the integral

$$\int_{R^3} |\tilde{g}_t(\lambda)|^2 \, d\lambda = \frac{1}{(2\pi)^3} \int_{R^3} \frac{|e^{i|t|\mu_1} - 1|^2}{|\mu|^4} \, d\mu =$$

$$= \int_{R^1} |e^{i|t|\mu_1} - 1|^2 \left[\frac{\mu_1^2}{\mu_1^4} \int_{R^2} \frac{d(\frac{\mu_2}{\mu_1})\, d(\frac{\mu_3}{\mu_1})}{[1 + (\frac{\mu_2}{\mu_1})^2 + (\frac{\mu_3}{\mu_1})^2]^2} \right] d\mu_1 =$$

$$= \frac{\sigma^2}{2\pi} \int_{R^1} \left| \frac{e^{i|t|\mu_1} - 1}{i\mu_1} \right|^2 d\mu_1 = \sigma^2 \|1_{(0,|t|)}\|^2_{L_2(R^1)} = \sigma^2 |t|,$$

$$\sigma^2 = 2\pi \int_{R^2} \frac{d\lambda}{(1 + |\lambda|^2)^2}.$$

We did not see yet that $X = \mathcal{L}^* L_2(T) = [D]$, $D = C_0^\infty(T)$. This can be verified with a help of the property (3.2), Chapter I, say, which follows from the integral representation (1.14) for $u = \varphi \in D$ with

$$|\varphi(t)| \le \|\tilde{g}_t\|_{L_2} \|\alpha\|_{L_2}$$

$$= \sigma\sqrt{t}\|\mathcal{L}\varphi\|_{L_2} \le C\|\mathcal{L}\varphi\|_{L_2}, \qquad t \in T_{\text{loc}}, \qquad \varphi \in D,$$

in any bounded region $T_{\text{loc}} \subseteq T$. ■

Let us consider equation (1.1) in the case of

$$\mathcal{L} = \sum_{|k|\le p} a_k \partial^k = \mathcal{L}(\partial)$$

with (real) constant coefficients

$$\mathcal{L}(\partial)u = f \tag{1.16}$$

in $T = R^d$, and apply our approach with $F = L_2(T)$, assuming that the polynomial

$$\mathcal{P}(i\lambda) = \mathcal{L}^*(i\lambda)\mathcal{L}(i\lambda) = |\mathcal{L}(i\lambda)|^2, \qquad \lambda \in R^d,$$

is such that $1/\mathcal{P}$ is locally integrable. Then our test functions space $X = \mathcal{L}^* L_2(R^d)$ is dual to the W-space with the norm

$$\|u\|_W = \|\mathcal{L}u\|_{L_2} = \|\tilde{u}\|_{L_{2,\mathcal{P}}}$$

given by means of Fourier transforms in the corresponding $L_{2,\mathcal{P}}$-space, and all $x \in X$ can be regarded as generalized functions with Fourier transforms $\tilde{x} \in L_{2,1/\mathcal{P}}$-space with the weight function $1/P$,

$$\widetilde{X} = L_{2,1/\mathcal{P}}(R^d)$$

— see (3.7), (3.8), Chapter I. As we know, in this case we have

$$D = C_0^\infty(R^d) \subseteq X, \qquad X = [D].$$

Of course the operator

$$\mathcal{L}^* = \sum (-1)^{|k|} a_k = \mathcal{L}(-\partial)$$

is nondegenerate, since for any $g \in L_2(R^d)$, $\mathcal{L}^* g = 0$, regarded as a generalized function, we have the generalized Fourier transform $\widetilde{\mathcal{L}^* g} = \mathcal{L}^*(i\lambda)\tilde{g}(\lambda) = 0$ only for $g = 0$. This again shows that $X = \mathcal{L}^* L_2(R^d)$ is the space of all generalized functions $x = \mathcal{L}^* g$, $g \in L_2(R^d)$, with

$$\tilde{x}(\lambda) = \mathcal{L}^*(i\lambda)\tilde{g}(\lambda) = \mathcal{L}(-i\lambda)\tilde{g}(\lambda)$$

and the norm

$$\|x\|_X = \|g\|_{L_2} = \|\tilde{x}\|_{L_{2,1/\mathcal{P}}} = \left(\int |\tilde{x}(\lambda)|^2 \frac{1}{\mathcal{P}}\, d\lambda\right)^{1/2}.$$

Having in mind the test functions $x \in D$ for $u \in \mathbf{W}$, we can apply the generalized Fourier transform $\tilde{f}$ of $f = \mathcal{L}u$ to get (x, u) as

$$(x, u) = (g, f) = (\tilde{g}, \tilde{f})$$

with

$$\tilde{g}(\lambda) = \tilde{x}(\lambda)/\mathcal{L}(-i\lambda), \qquad \lambda \in R^d.$$

It shows, for example, that in the case of the equation (1.16) with its right-hand side f represented by the white noise on the $L_2(R^d)$-space, we get the unique solution $u = \xi \in \mathbf{W}$ as a *generalized stationary random field* with the spectral density

$$\frac{1}{\mathcal{P}(i\lambda)} = \frac{1}{|\mathcal{L}(i\lambda)|^2}, \qquad \lambda \in R^d;$$

namely, according to the stochastic integral representation

$$(\tilde{\varphi}, \tilde{f}) = \int_{R^d} \tilde{\varphi}(\lambda)\, \tilde{\eta}(d\lambda), \qquad \varphi \in D,$$

of the white noise $\tilde{f}$ on the $L_2(R^d)$-space, we get for $u = \xi \in \mathbf{W}$ the corresponding *spectral representation*

$$(x, \xi) = \int_{R^d} \tilde{x}(\lambda) \frac{1}{\mathcal{L}(-i\lambda)}\, \tilde{\eta}(d\lambda), \qquad x \in D = C_0^\infty(R^d),$$

where the stochastic measure $\tilde{\eta}(d\lambda)$ with orthogonal (noncorrelated) values on R^d is such that

$$\mathrm{E}\tilde{\eta}(d\lambda) = 0, \qquad \mathrm{E}|\tilde{\eta}(d\lambda)|^2 = d\lambda,$$

and it gives

$$\mathrm{E}(x, \xi) = 0, \qquad \mathrm{E}|(x, \xi)|^2 = \int_{R^d} |\tilde{x}(\lambda)|^2 \frac{1}{\mathcal{P}(i\lambda)}\, d\lambda.$$

Assuming that $1/\mathcal{P}$ is integrable, i.e.,

$$\frac{1}{\mathcal{L}(i\lambda)} \in L_2(R^d),$$

we have in our test functions space X a complete system of delta-functions $x = \delta_t$ labelled by $t \in R^d$ because all $\tilde{x}(\lambda) = \frac{1}{(2\pi)^{d/2}} e^{-i\lambda t}$, $\lambda \in R^d$, form a complete system in the corresponding $\widetilde{X} = L_{2,1/\mathcal{P}}(R^d)$. Hence, any $u \in \mathbf{W}$ can be described as

$$u = u(t) = (\delta_t, u), \qquad t \in T = R^d,$$

where

$$(\delta_t, u) = (g_t, f), \qquad f = \mathcal{L}u,$$

with $\delta_t = \mathcal{L}^* g_t$,

$$g_t = g(t-s) = \frac{1}{(2\pi)^{d/2}} \int_{R^d} e^{-i\lambda(t-s)} \frac{1}{\mathcal{L}(-i\lambda)}\, d\lambda, \qquad s \in R^d.$$

And according to the stochastic integral representation

$$(\varphi, f) = \int_{R^d} \varphi(t)\, \eta(dt), \qquad \varphi \in D,$$

of the white noise f on the $L_2(R^d)$-space, we have $u = \xi \in \mathbf{W}$ as

$$\xi(t) = \int_{R^d} g(t-s)\, \eta(ds), \qquad t \in R^d. \quad ■$$

One can derive similar results in a case of the differential operator

$$\mathcal{L} = \sum a_k \partial^k = \mathcal{P}(\partial) \geq 0$$

with constant coefficients when the equation

$$\mathcal{P}(\partial) u = f \tag{1.17}$$

in $T = R^d$ is treated according to our general approach to $\mathcal{L} = \mathcal{L}^* = \mathcal{P} \geq 0$ in the W-type space $F = W = [C_0^\infty(T)]$ with $\|\varphi\|_W = (\varphi, P\varphi)^{1/2}$, $\varphi \in C_0^\infty(T)$. In particular, we have

$$\widetilde{X} = L_{2,1/\mathcal{P}}(R^d)$$

with

$$\mathcal{P}(i\lambda) = \sum a_k (i\lambda) \geq 0, \qquad \lambda \in R^d,$$

for our test functions space $X = \mathcal{L}^* F = \mathcal{P}W$. The unique solution $u \in \mathbf{W}$ of equation (1.17) can be obtained by means of the Fourier transform as

$$(x, u) = (g, f) = (\tilde{g}, \tilde{f}), \qquad x \in D,$$

with

$$\tilde{g}(\lambda) = \frac{\tilde{x}(\lambda)}{\mathcal{P}(i\lambda)}, \qquad \lambda \in R^d.$$

In the case of equation (1.17) with the right-hand side represented by the corresponding white noise on the space $F = W$, we have $u = \xi \in \mathbf{W}$ as the *generalized stationary random field* with *spectral density*

$$\frac{1}{\mathcal{P}(i\lambda)}, \qquad \lambda \in R^d,$$

namely,

$$\mathrm{E}|(x,\xi)|^2 = \|\tilde{x}\|^2_{L_{2,1/\mathcal{P}}} = \int\limits_{R^d} |\tilde{x}(\lambda)|^2 \frac{1}{\mathcal{P}(i\lambda)}\, d\lambda, \qquad x \in D.$$

Here as an example, we have the well-known *Markov Free Field* with the spectral density

$$\frac{1}{|\lambda|^2 + 1}, \qquad \lambda \in R^d,$$

which arises in our scheme from equation (1.17) with the differential operator

$$\mathcal{P} = -\Delta + 1$$

and the white noise f on the corresponding Sobolev space $F = W = W_2^1(R^d)$. ■

Let us consider a parabolic type evolution equation

$$\frac{\partial u}{\partial r} = Au + f \tag{1.18}$$

in $R^d = R^{d-1} \times R^1$ with $A = \sum_{|k| \le p} a_k \partial^k$ over a space variable $s \in R^{d-1}$ and $\partial/\partial r$ over a time variable $r \in R^1$ which represents an example of the equation (1.16) with

$$\mathcal{L} = \frac{\partial}{\partial r} - A$$

in the $L_2(R^d)$-space. Here we assumed that A is a symmetric elliptic operator in R^{d-1} such that in our general scheme of Section 3.1, Chapter I, the corresponding operator

$$\mathcal{P} = \mathcal{L}^* \mathcal{L} = -\frac{\partial^2}{\partial r^2} + A^2 \ge 0$$

on $D = C_0^\infty(R^d)$ leads to the Sobolev space

$$W = [D] = \overset{\circ}{W}{}_2^{(p,1)}(R^d)$$

with the norm

$$\|\varphi\|_W^2 = (\varphi, P\varphi)^{1/2} \asymp \|\varphi\|_{(p,1)}, \qquad \varphi \in D.$$

As we know (see Section 4.4, Chapter I) the dual Sobolev space $X = W^* = \overset{\circ}{W}{}_2^{-(p,1)}(R^d)$ which for equation (1.18) plays the role of the test functions space $X = \mathcal{L}^* L_2(R^d)$, contains a complete system of generalized test functions

$$x \times \delta_r, \qquad x \in C_0^\infty(R^{d-1}), \qquad r \in R^1,$$

and it let us test any $u \in \mathbf{W}$ as function of the space variable $s \in R^{d-1}$ by $x \in C_0^\infty(R^{d-1})$ and the time variable $r \in R^1$ by all delta-functions δ_r; as a result of this testing we obtain a representation of $u \in \mathbf{W}$ as

$$u = u_r, \qquad -\infty < r < \infty,$$

by means of its generalized trace

$$u_r = (x, u_r) = (x \times \delta_r, u), \qquad x \in C_0^\infty(R^{d-1}),$$

at $r \in R^1$ (cf. (5.11), Chapter I). Taking

$$g = x \times 1_{(a,b)}, \qquad x \in C_0^\infty(R^{d-1}), \qquad r \in R^1,$$

in the $L_2(R^d)$-space and the generalized test functions

$$x = \mathcal{L}^* g = x \times (\delta_b - \delta_a) - Ax \times 1_{(a,b)} \in X$$

we for any $u \in \mathbf{W}$ obtain

$$(\mathcal{L}^* g, u) = (x \times (\delta_b - \delta_a), u) - (Ax \times 1_{(a,b)}, u)$$

$$= [(x, u_b) - (x, u_a)] - \int_a^b (Ax, u_r)\, dr = (x \times 1_{(a,b)}, f), \qquad f = \mathcal{L} u.$$

In particular, for equation (1.18) with the white noise f on the $L_2(R^d)$-space, given by the known stochastic integral representation

$$(\varphi, f) = \int_{R^1} \varphi_r\, \eta_{dr}, \qquad \varphi \in D = C_0^\infty(R^d),$$

— cf. (1.9), Chapter I, we obtain for $u = \xi \in \mathbf{W}$ as $\xi = \xi_r$, $-\infty < r < \infty$, the corresponding stochastic integral equation

$$(x, \xi_b) - (x, \xi_a) = \int_a^b (Ax, \xi_r)\, dr + \int_a^b x\eta_{dr},$$

$$x \in C_0^\infty(R^{d-1}), \quad -\infty < a \leq b \leq \infty.$$

It can be treated as the generalized *stochastic Ito equation*

$$d\xi_r = A\xi_r\, dr + \eta_{dr} \tag{1.18)'$$

which is actually equivalent to the equation (1.18) for the generalized random function $\xi = u \in \mathbf{W}$,

$$(\varphi, u) = \int_{R^1} \varphi_r u_r\, dr, \qquad \varphi \in D = C_0^\infty(R^d),$$

and therefore, is equivalent also to the equation

$$-\frac{\partial^2}{\partial r^2}\xi + Q\xi = \eta$$

with $Q = A^2$ and the white noise η on the corresponding Sobolev space $W = \overset{\circ}{W}{}_2^{(p,1)}(R^d)$ — this new stochastic source η appears after applying the operator $\mathcal{L}^*$ to the former equation (1.18) with the white noise f on the $L_2(R^d)$-space. (Recall, the latter one was considered already — see (5.15), (5.16), Chapter I.)

II.2. Boundary conditions for generalized differential equations

2.1. Solvability of the equations

In our approach to the generalized differential equation (1.1), we introduced the proper test functions $x \in X$ which were employed to describe the unique solution $u \in \mathbf{W}$ in th region $T \subset R^d$.

It seems natural to employ the functional class $\mathbf{W}$ by considering

$$\mathcal{L}u = f \tag{2.1}$$

in the region $S \subseteq T$, and we encounter the problem to identify $u \in \mathbf{W}$ in S by equation (2.1) plus some appropriate *boundary conditions* on the boundary $\Gamma = \partial S$. What we have in mind is to identify the corresponding generalized (random) function

$$u = (\varphi, u) \quad \varphi \in C_0^\infty(S),$$

regarding it as $u \in \mathbf{W}(S)$ where the functional class $\mathbf{W}(S)$ is represented by *all* $u \in \mathbf{W}$ restricted in the region S.

But first we give the characterization of the generalized random functions $f = (\varphi, f)$, $\varphi \in C_0^\infty(S)$, which can play the role $f = \mathcal{L}u$, $u \in \mathbf{W}(S)$, in the region $S \subseteq T$.

This problem is of real interest only when we deal with the operator $\mathcal{L}$ in the space $F = L_2(T)$. For arbitrary $\mathcal{L} = \mathcal{L}^* = \mathcal{P} \geq 0$ in the corresponding $F = W$, all $f = \mathcal{L}u$ in the region $S \subseteq T$ can be generally characterized no better than in $S = T$ by continuity of (φ, f) with respect to $\|\varphi\|_W$, $\varphi \in C_0^\infty(S)$, in the meansquare sense. This characterization holds true in any region $S \subseteq T$ thanks to the fact that any linear continuous function $f = (\varphi, f)$, $\varphi \in C_0^\infty(S)$, can be extended on the subspace

$$F(S) = [C_0^\infty(S)] \subseteq W = F$$

and then, on its direct complement, to $f = (\varphi, f)$, $\varphi \in F = W$.

Let us recall that for the operator $\mathcal{L}$ in the space $F = L_2(T)$ all generalized random functions $f = \mathcal{L}u$, $u \in \mathbf{W}$, in the very region T are characterized by the meansquare continuity of $(\varphi, f) = (\mathring{\varphi}, f)$ with respect to $\| \mathring{\varphi} \|_{L_2} = \|\mathcal{L}^*\varphi\|_X$, $\varphi \in C_0^\infty(T)$, where $\mathring{\varphi}$ is a projection of φ onto the subspace $\mathring{F} = [\mathcal{L}C_0^\infty(T)]$ in $F = L_2(T)$. Obviously, this characterization holds true in any region $S \subseteq T$ as well, for every $f = (\mathring{\varphi}, f)$, $\varphi \in C_0^\infty(S)$, can be extended to the meansquare linear continuous function $f = (\mathring{\varphi}, f)$, $\varphi \in L_2(T)$, which gives the corresponding $f = (\varphi, f)$, $\varphi \in C_0^\infty(T)$, with $(\varphi, f) \equiv (\mathring{\varphi}, f)$. This answers our question about possible $f = \mathcal{L}u$ in the case when we deal with the *nondegenerate* operator $\mathcal{L}^*: L_2(T) \to X$ and we have $\mathring{F} = L_2(T)$, $\varphi = \mathring{\varphi}$. It is worth mentioning here that for the often encountered case of $\mathcal{L}^*$ in the region S with the closure $[S] \subseteq T$, we have the equivalence

$$\|\mathcal{L}^*\varphi\|_X \equiv \| \mathring{\varphi} \|_{L_2} \asymp \|\varphi\|_{L_2}, \qquad \varphi \in L_2(S). \tag{2.2}$$

Therefore, in this case all generalized random functions $f = \mathcal{L}u$, $u \in \mathbf{W}$, in the region S are completely characterized by continuity of (φ, f) in the meansquare sense with respect to $\|\varphi\|_{L_2}$, $\varphi \in C_0^\infty(S)$, and can be considered as meansquare linear continuous functions

$$f = (\varphi, f), \qquad \varphi \in F(S) = L_2(S).$$

The fact that $\mathcal{L}^*$ is of type (2.2), can be verified by the solvability of equation (2.1) for any *deterministic* function $f \in L_2(S)$. Namely, for deterministic function $f = \mathcal{L}u$ with $u \in \mathbf{W}(S)$ we have

$$(\varphi, f) = (\mathcal{L}^*\varphi, u), \qquad \varphi \in C_0^\infty(S),$$

continuous with respect to $\|\mathcal{L}^*\varphi\|_X = \| \overset{\circ}{\varphi} \|_{L_2}$ and $|(\varphi, f)|$ is bounded on a set of all $\varphi \in C_0^\infty(S)$, $\| \overset{\circ}{\varphi} \|_{L_2} \leq 1$. If this holds true for any $f \in L_2(S)$, i.e., the set of $\varphi \in C_0^\infty(S)$: $\| \overset{\circ}{\varphi} \|_{L_2} \leq 1$ is weakly bounded in the L_2-space, then $\|\varphi\|_{L_2} \leq C$; hence

$$\| \overset{\circ}{\varphi} \|_{L_2} \leq \|\varphi\|_{L_2} \leq C \| \overset{\circ}{\varphi} \|_{L_2}, \qquad \varphi \in C_0^\infty(S). \tag{2.2$'$}$$

It is easy to see that existence of $u \in \mathbf{W}(S)$ with its *deterministic* part $\mathcal{L}u = f \in L_2(S)$ implies existence of *deterministic* function $u \in W(S)$, where $W(S)$ represents a restriction of the corresponding space $W = X^*$ in the region S.

Thus, we can characterize the operator $\mathcal{L}^*$ appealing to the deterministic Partial Differential Equations as follows.

LEMMA. *The operator $\mathcal{L}^*$ in the L_2-space is of the type* (2.2) *if and only if the deterministic equation* (2.1) *has the solution $u \in W(S)$ for any $f \in L_2(S)$.*

2.2. Boundary test functions

Dealing with the restriction of $u \in \mathbf{W}$ in the region $S \subseteq T$ as a generalized random function

$$u = (x, u), \qquad x \in C_0^\infty(S),$$

we have this $u \in \mathbf{W}(S)$ well-defined on all test functions $x \in X(S)$ which arise as a limit of $x \in C_0^\infty(S)$ in our test functions space X, i.e., we have the corresponding space

$$X(S) = [C_0^\infty(S)] \subseteq X$$

of generalized test functions x, $\operatorname{supp} x \subseteq [S]$, with supports in the closure of the region S, to serve $u \in \mathbf{W}(S)$ as

$$u = (x, u), \qquad x \in X(S). \tag{2.3}$$

In particular, for $\mathcal{L}^* C_0^\infty(S) \subseteq C_0^\infty(S)$ serving equation (2.1) in the region S, we have

$$\mathcal{L}^* C_0^\infty(S) \subseteq \mathcal{L}^* F(S) \subseteq [\mathcal{L}^* C_0^\infty(S)] \subseteq X(S)$$

in $X(S)$ where $F(S) = [C_0^\infty(S)]$ is a closure of $C_0^\infty(S)$ in the corresponding Hilbert space F. Let us recall that $\|\mathcal{L}^*\varphi\|_X \leq \|\varphi\|_F$ and in the known case $\|\mathcal{L}^*\varphi\|_X \asymp \|\varphi\|_F$, $\varphi \in C_0^\infty(S)$, we have $\mathcal{L}^* F(S) = [\mathcal{L}^* C_0^\infty(S)]$ as a closure of $\mathcal{L}^* C_0^\infty(S)$ which turns equation (2.1) into

$$(\mathcal{L}^* g, u) = (g, f), \qquad f \in F(S) = [C_0^\infty(S)], \tag{2.1$'$}$$

— cf. (1.1)$'$, (1.2).

One can observe that for any function $u \in \mathbf{W}(S)$ treated as suggested in (2.3), equation (2.1), (2.1)$'$ with given $f = \mathcal{L}u$ in the region S describes a part

$$u = (x, u), \qquad x \in X^-(S), \tag{2.3$'$}$$

of $u \in \mathbf{W}(S)$ on the subspace

$$X^-(S) = [\mathcal{L}^* C_0^\infty(S)] \subseteq X(S).$$

It is worth mentioning again that in our scheme for any operator $\mathcal{L}$ with $\mathcal{L}^*$ of the type (2.2) on

$$F(S) = [C_0^\infty(S)] = L_2(S)$$

or any operator $\mathcal{L} = \mathcal{L}^* = \mathcal{P} \geq 0$ on the corresponding

$$F(S) = [C_0^\infty(S)] = W(S)$$

we have

$$X^-(S) = [\mathcal{L}^* C_0^\infty(S)] = \mathcal{L}^* F(S).$$

A remarkable thing is that a complement to $X^-(S)$ in $X(S) = [C_0^\infty(S)]$ can be chosen as an appropriate collection $X^+(\Gamma)$ of *boundary test functions*

$$x \in X(S), \qquad \operatorname{supp} x \subseteq \Gamma,$$

with supports on the boundary $\Gamma = \partial S$. (For example, $X^+(\Gamma) = X_0^+(\Gamma)$ of this kind can be chosen as the orthogonal complement $X_0^+(\Gamma) = X(S) \ominus X^-(S)$ to $X^-(S)$ in $X(S)$.) And equation (2.1), (2.1)′, giving (2.3)′, alltogether with the corresponding *generalized boundary values*

$$(x, u), \quad x \in X^+(\Gamma) \tag{2.3''}$$

describe completely $u \in \mathbf{W}(S)$ in the region $S \subseteq T$ as in (2.3). To show this for $X^+(\Gamma) = X_0^+(\Gamma)$, we apply our general scheme of Section 3.1, Chapter I, with the corresponding $P \geq 0$ and all $x \in X$ regarded as generalized functions of the type

$$(\varphi, x) = \langle \mathcal{P}\varphi, x\rangle_X, \qquad \varphi \in D = C_0^\infty(T).$$

In the case of the differential operator $\mathcal{L}$ in the space $F = L_2(T)$ with $P = \mathcal{L}^*\mathcal{L}$ and

$$X = PW = \mathcal{L}^* L_2(T), \qquad W = X^*,$$

we have

$$X^-(S) = [\mathcal{L}^* C_0^\infty(S)] \supseteq [\mathcal{L}^*\mathcal{L} C_0^\infty(S)] = [\mathcal{P} C_0^\infty(S)].$$

Hence

$$X_0^+(\Gamma) = X(S) \ominus X^-(S) \subseteq X(S) \ominus [\mathcal{P} C_0^\infty(S)] = X(\Gamma),$$

where the *boundary test functions space*

$$X(\Gamma) = X(S) \ominus [\mathcal{P} C_0^\infty(S)] \tag{2.4}$$

represents all $x \in X(S)$, $\operatorname{supp} x \subseteq \Gamma$, with supports on the boundary $\Gamma = \partial S$, since the generalized functions $x \in X(\Gamma)$ with supports in the closure $[S]$ are characterized by the property that they vanish in the region S according to

$$(\varphi, x) = \langle \mathcal{P}\varphi, x\rangle_X = 0, \qquad \varphi \in C_0^\infty(S).$$

Moreover, for the operator $\mathcal{L}$ in $F = L_2(T)$, we have

$$X^-(\Gamma) = [\mathcal{L}^* C_0^\infty(S)] \ominus [\mathcal{P} C_0^\infty(S)] = [\mathcal{L}^* C_0^\infty(S)] \cap X(\Gamma) \tag{2.5}$$

representing the orthogonal complement to $X_0^+(\Gamma) \subseteq X(\Gamma)$ in the boundary test functions space $X(\Gamma) \subseteq X(S)$,

$$X_0^+(\Gamma) = X(\Gamma) \ominus X^-(\Gamma) \tag{2.6}$$

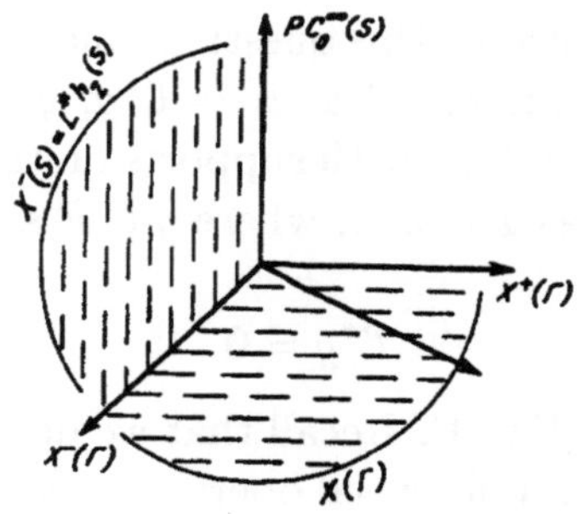

Fig. 1

— see Fig. 1.

In the case of the differential operator $\mathcal{L} = \mathcal{L}^* = \mathcal{P} \geq 0$ with corresponding $F = W$, $X = \mathcal{P}W$, we have

$$X^-(S) = [\mathcal{P}C_0^\infty(S)]$$

From the general representation (2.4), we see at once that

$$X_0^+(\Gamma) = X(\Gamma) \tag{2.7}$$

represents all boundary test functions $x \in X(S)$, $\operatorname{supp} x \subseteq \Gamma = \partial S$.

In both cases considered above, we have our test functions space $X(S)$ in the region S in the form of the direct sum

$$X(S) = X^-(S) + X^+(\Gamma) \tag{2.8}$$

where $X^-(S) = [\mathcal{L}^* C_0^\infty(S)]$ serving the very equation (2.1), (2.1)$'$ gives us part (2.3)$'$ of $u \in \mathbf{W}(S)$, and $X^+(\Gamma) = X_0^+(\Gamma)$ serving its additional part gives the corresponding boundary values (2.3)$''$. Hence, we obtain $u \in W(S)$ as

$$u = (x, u) = (x^-, u) + (x^+, u)$$

with the decomposition $x = x^- + x^+$ of $x \in X(S) = [C_0^\infty(S)]$ on the components $x^- \in X^-(S)$ and $x^+ \in X^+(\Gamma)$. ■

What kind of boundary test functions $x \in X(\Gamma)$ we actually have at our disposal? Concerning this question, one can in general observe that any $x \in X$ is represented as $x = \mathcal{L}^* g$, $g \in F$, and boundary test functions x, $\operatorname{supp} x \subseteq \Gamma$, with supports on the boundary $\Gamma = \partial S$ can be characterized as $x = \mathcal{L}^* g$, where $g \in F$ satisfies a generalized differential equation

$$\mathcal{L}^* g = 0 \tag{2.9}$$

in the complement $T \backslash \Gamma$ to Γ. Recall that we deal with $F = L_2(T)$ in the case of the general differential operator $\mathcal{L}$ and the corresponding $F = W$ in the case of $\mathcal{L} = \mathcal{L}^* = \mathcal{P} \geq 0$.

In particular, in the case of $\mathcal{L}$ in the space $F = L_2(T)$, having $X = \mathcal{L}^* L_2(T)$, we can "slice" $g \in L_2(T)$ into two parts as $g = g^- \oplus g^+$ with $g^- \in L_2(S)$ in S and $g^+ \in L_2(S^c)$ in the complement $S^c = T \backslash S$. And for $g = g^-$, g^+ which satisfy equation (2.9) we have correspondingly all $x = x^- = \mathcal{L}^* g^-$ representing $X^-(\Gamma) \subseteq X^-(S)$, and all $x = x^+ = \mathcal{L}^* g^+$ with $g^+ = 0$ in the very region S,

$$\mathcal{L}^* g^+ = 0$$

in the complementary region $S^+ = T \backslash [S]$ representing a particular complement $X^+(\Gamma) \subseteq X(\Gamma)$ to $X^-(\Gamma)$ in $X(\Gamma)$. Here for $x = x^+$ the boundary values (x, u) of $u \in \mathbf{W}(S)$ can be determined by means just of $f = \mathcal{L} u$ outside S as

$$(x, u) = (g^+, f), \qquad x = \mathcal{L}^* g^+ \in X^+(\Gamma),$$

where $X^+(\Gamma)$, in the case of the nondegenerate operator $\mathcal{L}^*$, serving as the unitary operator $\mathcal{L}^* \colon L_2(T) \to X$, represents the orthogonal complement

$$X^+(\Gamma) = X_0^+(\Gamma) = X(S) \ominus X^-(S).$$

If there are no solutions $g \in L_2(S)$, $g \not\equiv 0$, of equation (2.9), then $X^-(\Gamma) = 0$ and

$$X^+(\Gamma) = X_0^+(\Gamma) = X(\Gamma)$$

(see (2.5) and (2.6)).

There is a question whether any $x \in X$, $\operatorname{supp} x \subseteq \Gamma$, belongs to $X(\Gamma)$, i.e., $x \in X(S) = [C_0^\infty(S)]$, as it holds true for Sobolev spaces — see Section 4.2, Chapter I.

2.3. General boundary conditions

As was defined, the generalized random functions $u \in \mathbf{W}$ in the region S are characterized by meansquare continuity of (φ, u), $\varphi \in C_0^\infty(S)$, with respect to the corresponding norm $\|\varphi\|_X$, and all generalized functions of this kind can uniquely be extended on the closure $X(S) = [C_0^\infty(S)]$ in X as $u = (x, u)$, $x \in X(S)$; recall that any $u \in \mathbf{W}(S)$ is actually a part of some $u \in \mathbf{W}$: $u = (x, u)$, $x \in X$. Having this in mind, we consider various boundary conditions for $u \in \mathbf{W}(S)$ on the boundary $\Gamma = \partial S$, by choosing an appropriate collection $x \in X^+(\Gamma)$ of *boundary test functions* $x \in X$, $\operatorname{supp} x \subseteq \Gamma$, and setting

$$(x, u) = (x, u^+), \qquad x \in X^+(\Gamma), \tag{2.10}$$

by means of some *sample* system of $u^+ \in \mathbf{W}(S)$.

(Note, in what follows, with no change of notations, one can consider $u \in \mathbf{W}(S)$ extended on the closed region $[S]$ as $u = (x, u)$, $x \in X(S)$, where $X(S)$ is a collection of all $x \in X$, $\operatorname{supp} x \subseteq [S]$, with supports in $[S]$. Then the corresponding boundary test functions space $X(\Gamma) \subseteq X(S)$ is formed by all $x \in X$, $\operatorname{supp} x \subseteq \Gamma \in [S]$. In the framework of this scheme, one can safely apply the description of $x \in X(\Gamma)$ given in (2.9).)

Any chosen collection $X^+(\Gamma) \subseteq X(\Gamma)$ determines the corresponding *type* of the boundary conditions (2.10) given with the appropriate sample $u^+ \in \mathbf{W}(S)$ which allow us to prescribe the boundary values (x, u), $x \in X^+(\Gamma)$.

Of course, we have the same type of boundary conditions (2.10) for any collection $X^+(\Gamma)$ with the same linear closure $[X^+(\Gamma)]$. Since $u = (x, u)$ are meansquare *linear* continuous functions of $x \in X$, it is convenient to regard $X^+(\Gamma)$ as a linear subspace in the boundary test functions space $X(\Gamma)$, setting the boundary conditions (2.10) by means of an appropriate *complete system* of $x \in X^+(\Gamma)$. ■

Our general approach to boundary conditions with the appropriate subspace $X^+(\Gamma) \subseteq X(\Gamma)$ of boundary test functions can be exposed as follows for *deterministic* $u \in W(S)$ in the corresponding functional class $W(S)$ which is a restriction of

$$W = [C_0^\infty(T)] = X^*$$

in the region $S \subseteq T$ (to be more precise, $W(S) = X(S)^*$). We have in mind the boundary conditions on $\Gamma = \partial S$ associated with some *boundary operator* γ on $u \in W(S)$,

$$\gamma(\varphi) = 0, \qquad \varphi \in C_0^\infty(S) \subseteq W(S),$$

and set by means of an appropriate *sample* $u^+ \in W(S)$ as

$$\gamma(u - u^+) = 0 \tag{2.11}$$

Here, for a closed subspace of all φ: $\gamma(\varphi) = 0$, $\varphi \in W(S)$, which is uniquely characterized by its *annulator* $X^+(\Gamma) \subseteq X(S)$ in the dual space $X(S) = W(S)^*$, this kind of boundary condition can be described exactly as

$$(x, u - u^+) = 0, \qquad x \in X^+(\Gamma),$$

And it is what we have in (2.10), since all $x \in X^+(\Gamma)$ have their supports $\operatorname{supp} x \subseteq \Gamma$ on the boundary $\Gamma = \partial S$ according to

$$(x, \varphi) \equiv (\varphi, x) = 0, \qquad \varphi \in C_0^\infty(S),$$

so the annulator $X^+(\Gamma) \subseteq W(S)^*$ represents the certain boundary subspace in $X(\Gamma)$.

For illustration, let us consider our scheme with the elliptic differential operator $\mathcal{L} = \sum_{|k|\leq p} a_k \partial^k$ in $F = L_2(T)$ when we have $W = X^*$ for our test functions space $X = \mathcal{L}^* L_2(T)$ with the norm

$$\|\varphi\|_W \asymp \|\varphi\|_p, \qquad \varphi \in C_0^\infty(T_{\text{loc}}),$$

in any bounded region $T_{\text{loc}} \subseteq T$. Then, for the bounded region S, $[S] \subseteq T$, we have the corresponding Sobolev spaces

$$W(S) = W_2^p(S), \qquad X(S) = W_2^{-p}(S).$$

And with what we know about the boundary test functions space $X(\Gamma)$ on the boundary $\Gamma = \partial S$, we can set the boundary condition of the type (2.11) by means of any operator γ,

$$\gamma(u) = C(u^{(0)}, \ldots, u^{(p-1)}),$$

with a constant matrix C, say, applied to the boundary trace

$$(u^{(0)}, \ldots, u^{(p-1)}) \in \prod_{k=0}^{\infty} \mathring{W}_2^{p-k-1/2}(\Gamma)$$

of $u \in W(S)$ and it's nontangent derivatives on Γ (see (4.12)–(4.15), Chapter I). In particualr, considering the Laplace operator $\mathcal{L} = \Delta$ in $F = L_2(T)$ leading to the Sobolev spaces

$$W(S) = W_2^2(S), \qquad X(S) = W_2^{-2}(S)$$

in the bounded region S, $[S] \subseteq T$, we can set, as in (2.11), the known *Dirichlet condition* with

$$\gamma(u) = u^{(0)},$$

or the *Neuman condition* with

$$\gamma(u) = u^{(1)},$$

etc.[1] ■

As has been shown above, in the framework of our scheme with the operator $\mathcal{L}$ in the space $F = L_2(T)$, there are certain boundary values

$$(x, u) = (g, f), \qquad x = \mathcal{L}^* g \in X^-(\Gamma)$$

$g \in L_2(S)$, $g = 0$ outside S and

$$\mathcal{L}^* g = 0 \tag{2.12}$$

in the region S, which are determined by the very equation (2.1), (2.1)′ — see (2.5). Therefore, general boundary conditions (2.10) for $u \in \mathbf{W}(S)$ with $f = \mathcal{L}u$ in S must be given in a way that

$$(x, u^+) = (g, f), \qquad x = \mathcal{L}^* g, \tag{2.13}$$

for $x \in X^+(\Gamma)$ which are in $X^-(\Gamma)$. Moreover, in the case of any boundary test functions $x \in X^+(\Gamma)$ such that $x - \mathcal{L}^* g \to 0$ for some $\mathcal{L}^* g \in X^-(\Gamma)$, we must necessarily have

$$(x, u^+) - (g, f) \to 0. \tag{2.13$'$}$$

Relations (2.13) and (2.13)′ express *consistency* of the boundary conditions (2.10) on $\Gamma = \partial S$ with equation (2.1)/(2.1)′ in the region S.

[1]Note, that in this very way elliptic boundary problems associated with the Sobolev spaces and the corresponding embeddings theorems were considered by S.L. Sobolev in this famous work "Some applications of functional analysis in mathematical physics," Leningrad University, 1950.

Recall, in our framework we have the same test functions space $X(S) = [C_0^\infty(S)]$ in X either for the general operator $\mathcal{L}$ in $F = L_2(T)$ or the symmetric positive operator $\mathcal{P} = \mathcal{L}^*\mathcal{L}$ in the corresponding $F = W$. Acordingly, for $u \in \mathbf{W}(S)$ the equation (2.1) with the boundary conditions (2.10) can be equivalently transfered to the equation

$$\mathcal{P}u = \mathcal{L}^* f$$

with the enlarged boundary conditions

$$(x, u) = \begin{cases} (x, u^+) = (g, f), & x = \mathcal{L}^* g \in X^-(\Gamma), \\ (x, u^+), & x \in X^+(\Gamma), \end{cases}$$

so the consistency of (2.10) with (2.1) simply means that these enlarged boundary conditions are not overdetermined. ■

Let us consider our scheme with $\mathcal{L}$ in $F = L_2(T)$ and boundary conditions (2.10) of a type such that the orthogonal complement

$$X^{++}(\Gamma) = X^+(\Gamma) \ominus X^{-+}(\Gamma)$$

to all

$$x \in X^-(\Gamma) \cap X^+(\Gamma) = X^{-+}(\Gamma)$$

in (2.13) plus $X^-(\Gamma)$ give us a direct sum

$$X^-(\Gamma) + X^{++}(\Gamma). \tag{2.14}$$

Then, along with the orthogonal complement

$$X^{--}(\Gamma) = X^-(\Gamma) \ominus X^{-+}(\Gamma)$$

we have a direct sum

$$X^{--}(\Gamma) + X^{-+}(\Gamma) + X^{++}(\Gamma)$$

representing a linear closure of $X^-(\Gamma)$ plus $X^+(\Gamma)$. Indeed, for any elements $x^{--} + x^{-+} + x^{++} \to 0$ with $x^{--} \in X^{--}(\Gamma)$, $x^{-+} \in X^{-+}(\Gamma)$, $x^{++} \in X^{++}(\Gamma)$ we have $x^{-+} \to 0$ as their orthogonal projections on $X^{-+}(\Gamma)$; therefore, $x^{--} + x^{++} \to 0$ and in the case of the direct sum in (2.14) it could happen only if $x^{--} \to 0$, $x^{++} \to 0$. According to (2.4)–(2.6), with the orthogonal complement

$$X^{--}(S) = X^-(S) \ominus X^{-+}(\Gamma)$$

to $X^{-+}(\Gamma)$ in $X^{-}(S) = \mathcal{L}^* L_2(S)$ and, if necessary, a properly enlarged $X^{++}(\Gamma) \subseteq X(\Gamma)$, we can obtain a direct sum

$$X^{--}(S) + X^{-+}(\Gamma) + X^{++}(\Gamma) = X(S) \tag{2.15}$$

representing our test functions space $X(S)$ in the region S.

Obviously, having the solvable equation (2.1) and the consistency condition (2.13) for all $x = \mathcal{L}^* g \in X^{-+}(\Gamma)$, from the formula

$$(x, u) = \begin{cases} (g, f), & x = \mathcal{L}^* g \in X^{-}(S), \\ (x, u^{+}), & x \in X^{+}(\Gamma), \end{cases} \tag{2.16}$$

we can get a solution $u \in \mathbf{W}(S)$ of the boundary problem (2.1), (2.10). Indeed, with application of the proper direct sum (2.15), for all $x \in X(S)$ as $x = x^{--} + x^{-+} + x^{--}$ with components $x^{--} \in X^{--}(S)$, $x^{-+} \in X^{-+}(\Gamma)$, $x^{++} \in X^{++}(\Gamma)$ we get the corresponding

$$u = (x, u) = (x^{--}, u) + (x^{-+}, u) + (x^{++}, u), \qquad x \in X(S).$$

Of course, in the particular case of $X^{-+}(\Gamma) = 0$, $X^{++}(\Gamma) = X^{+}(\Gamma)$, we just have the direct sum

$$X^{-}(\Gamma) + X^{+}(\Gamma) \tag{2.17}$$

and with $X^{+}(\Gamma)$ properly enlarged, if necessary, we obtain the direct sum

$$X^{-}(S) + X^{+}(\Gamma) = X(S) \tag{2.18}$$

— see Fig. 2 illustrated formula (2.16) which in this case gives

$$(\varphi, u) = (g, f) + (x, u^{+}), \qquad \varphi = \mathcal{L}^* g + x, \tag{2.16'}$$

for all $\varphi \in X(S) = [C_0^\infty(S)]$.

The consistency of the boundary conditions (2.10) of the general type (2.14), (2.15) is given by (2.13) alone, with no reference to (2.13)′.

One can verify that $X^{+}(\Gamma)$ with given collection $X^{-+} \subseteq X^{+}(\Gamma)$ of some $x = \mathcal{L}^* g$, $g \in L_2(S)$, determines the type (2.14), (2.15) as follows.

LEMMA. *Suppose that the deterministic equation* (2.1) *with arbitrary* $f \in L_2(S)$ *such that*

$$(g, f) = 0, \qquad g \in L_2(S), \qquad \mathcal{L}^* g \in X^{-+}$$

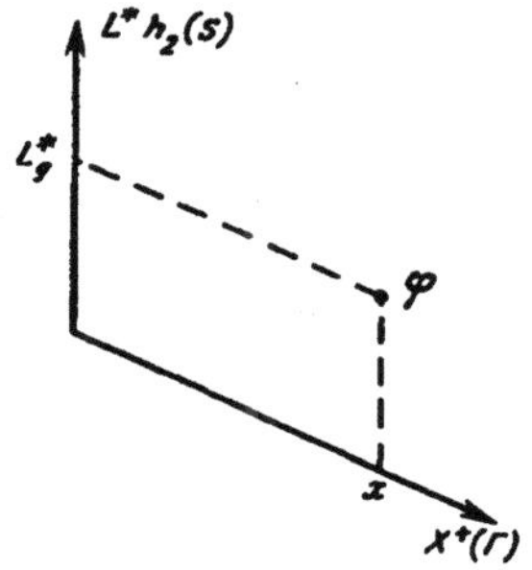

Fig. 2

has the solution $u \in W(S)$ *with the zero boundary conditions* (2.10) *on* $X^{+}(\Gamma) \supseteq X^{-+}(\Gamma)$. *Then,* $X^{+}(\Gamma)$ *is of the type* (2.14), (2.15) *with* $X^{-+}(\Gamma)$ *as a linear closure of the given* $X^{-+}(\Gamma)$.

Indeed, in the deterministic case, for any $u \in W(S)$ we have $f = \mathcal{L}u \in L_2(S)$. The solvability of equation (2.1) with the zero boundary conditions (2.10) for arbitrary $f \in L_2(S)$ is equivalent to the existence of a linear continuous functional $u \in W(S) = X(S)^*$, which on the orthogonal complement

$$X^{--}(S) = X^{-}(S) \ominus X^{-+}(\Gamma)$$

is arbitrary given in terms of $f = \mathcal{L}u$ as $(\mathcal{L}^* g, u) = (g, f)$, $g \in L_2(S)$, and is zero on the orthogonal complement

$$X^{++}(\Gamma) = X^{+}(\Gamma) \ominus X^{-+}(\Gamma),$$

and it does exist only when we have the direct sum (2.14), (2.15)).

The same result can be obtained in a slightly different way.

LEMMA. *Suppose that there exists a solution* $u \in W(S)$ *of the homogeneous equation* (2.1) *with* $f = 0$ *and the arbitrary boundary conditions* (2.10) *with any deterministic* $u^{+} \in W(S)$,

$$(x, u^{+}) = 0, \qquad x \in X^{-+}.$$

Then $X^{+}(\Gamma)$ *is of the type* (2.14), (2.15).

To explain this, let us consider a general case when any linear continuous functional on a subspace X_1 in a Hilbert space can be extended on a linear closure of X_1, X_2 with zero values on X_2. In this case any such functional $u = (x, u)$, $x = x_1 + x_2$, is bounded on a set $\{x_1\}$ of all $x_1 \in X_1$ with the corresponding appropriate elements $x_2 \in X_2$ which make $\|x_1 + x_2\| \leq 1$, namely,

$$|(x_1, u)| = |(x_1 + x_2, u)| \leq C\|x_1 + x_2\| \leq C.$$

Hence, the weakly bounded set $\{x_1\} \subseteq X_1$ is strongly bounded, and this shows that

$$\|x_1\| \leq C\|x_1 + x_2\|$$

for all $x_1 \in X_1$, $x_2 \in X_2$ which means that we have a direct sum $X_1 + X_2$. ■

Let us call the boundary conditions (2.10) *complete* if $x \in X^+(\Gamma)$ together with $x \in X^-(\Gamma)$ form a complete system in the boundary space $X(\Gamma)$. According to (2.4), (2.5) it is the same as that $x \in X^+(\Gamma)$ together with $x \in X^-(S)$ form a complete system in the whole test functions space $X(S)$ for $u \in \mathbf{W}(S)$. (Of course, a solution of equation (2.1) is not unique when the boundary conditions (2.10) are not complete.)

One can apply the following obvious proposition.

LEMMA. *Suppose that the homogeneous equation* (2.1) *with* $f = 0$ *and the zero boundary conditions* (2.10) *has the only solution* $u = 0$ *in* $W(S)$. *Then these boundary conditions are complete.*

Let us briefly formulate the results which can be obtained in the framework of our general scheme with the operator $\mathcal{L}$ in the space $F = L_2(T)$.

THEOREM. *The boundary problem* (2.1), (2.10) *does have a unique solution* $u \in \mathbf{W}(S)$ *if and only if equation* (2.1) *is solvable and the boundary conditions* (2.10) *are complete and consistent with this equation. The solution then can be described by the formula* (2.16).

Any $x \in X(S)$ is a limit of $x_n = x_n^- + x_n^+$ with $x_n^- \in X^-(S)$, $x_n^+ \in X^+(\Gamma)$ and

$$(x, u) = \lim(x_n, u) = \lim[(x_n^-, u) + (x_n^+, u)]$$

is well-defined because

$$[(x_n^- - y_m^-, u) + (x_n^+ - y_m^+, u)] \to 0$$

for any $y_m = y_m^- + y_m^+ \to x$ thanks to the consistency condition with

$$x_n - y_m = [(x_n^- - y_m^-) + (x_n^+ - y_m^+)] \to 0, \qquad n, m \to \infty.$$

And $(x, u) \to 0$ for $x \to 0$. Indeed, it holds for all $x = x^- + x^+$ with $x^- \in X^-(S)$, $x^+ \in X^+(\Gamma)$, since (2.1), (2.10) give $(x^-, u) \to 0$, $(x^+, u) \to 0$ for $x^- \to 0$, $x^+ \to 0$.

Of course, in the case of the *direct sum* representation $x = x^- + x^+$, which is characterized by the property

$$\|x^-\|_X + \|x^+\|_X \le C\|x^- + x^+\|_X$$

we do not need a consistency condition, because we have

$$\|(x^- + x^+, u)\| \le \|(x^-, u)\| + \|(x^+, u)\| \le$$

$$\le C_1\|x^-\|_X + C_2\|x^+\|_X \le C\|x^- + x^+\|_X. \quad \blacksquare$$

With probability point of view, boundary problem (2.1), (2.10) seems most interesting in a case when the stochastic source in equation (2.1) can be like white noise, say, and the stochastic boundary conditions (2.10) can be set independently on this source.

Taking it in account, in our scheme with the differential operator $\mathcal{L}$ in $F = L_2(T)$ we employ from *deterministic* PDE a case when *deterministic equation* (2.1) *with any right-hand side* $f \in L_2(S)$ *and zero boundary conditions* (2.10) *of the given type* (2.11) *has a unique solution* $u \in W(S)$, or when *deterministic equation* (2.1) *with zero right-hand side and any sample* $u^+ \in W(S)$ *in boundary conditions* (2.11) *has a unique solution* $u \in W(S)$. This gives us some kind of *comparison criteria* for the existence of solution we are looking for[2], thanks to our lemmas, and implies the following result.

THEOREM. *A unique solution* $u \in \mathbf{W}(S)$ *of the stochastic boundary problem* (2.1), (2.10) *does exist for any stochastic source* $f = (\varphi, f)$, $\varphi \in C_0^\infty(S)$, *meansquare continuous with respect to* $\|\varphi\|_F$, *and any stochastic sample* $u^+ \in \mathbf{W}(S)$ *in boundary conditions* (2.10).

Recall, the boundary problem (2.1), (2.10) for the general differential operator $\mathcal{L}$ in $F = L_2(T)$ can be equivalently transfered to

[2]In a variety of books on deterministic PDE see, for example, O.A. Ladygenskaya, Boundary Problems for Mathematical Physics Equations, Moscow, Nauka, 1973; J.L. Lions and E. Magenes, Nonhomogeneous Boundary Value Problems and Applications, Berlin, Heidelberg, New York, Springer, 1972.

the certain boundary problem for the operator $\mathcal{P} = \mathcal{L}^*\mathcal{L}$ in the corresponding $F = W$. And in our scheme for the general symmetric positive operator $\mathcal{L} = \mathcal{L}^* = \mathcal{P} \geq 0$ in $F = W$ we have the only type of *complete* boundary conditions with

$$X^+(\Gamma) = X(\Gamma)$$

— cf. with the Dirichlet type boundary conditions in (5.12), (5.13) and (5.15), (5.16), Chapter I. In this case the following result holds.

THEOREM. *A solution* $u \in \mathbf{W}(S)$ *of the stochastic boundary problem* (2.1), (2.10) *does exist for any stochastic source* $f = (\varphi, f)$, $\varphi \in C_0^\infty(S)$, *meansquare continuous with respect to* $\|\varphi\|_W$, *and any boundary conditions with the stochastic sample* $u^+ \in \mathbf{W}(S)$.

2.4. Stochastic generalized ordinary differential equations

Mostly for illustration, let us consider our general scheme with the ordinary differential operator

$$\mathcal{L} = \sum_{k=0}^{p} a_k \frac{d^k}{dt^k} \qquad (a_p = 1)$$

in the space $F = L_2(T)$ on $T = (0, \infty)$, say.

Recall, any classical solution of the equation $\mathcal{L}\varphi = \psi$ on the interval $T = (0, \infty)$ with initial values $\varphi^{(k)}(0) = 0$, $k = 0, \ldots, p$, can be represented as

$$\varphi(t) = \int_0^t g(t,s)\psi(s)\, ds, \qquad t \geq 0,$$

by means of the so-called resolvent $g(t,s)$, $t \geq s$, representing a solution of $\mathcal{L}g = 0$ on any interval (s, ∞) with initial values

$$\frac{d^k}{dt^k} g(t,s)\Big|_{t=s} = \begin{cases} 0, & k < p-1 \\ 1, & k = p-1. \end{cases}$$

One can see it at once, employing

$$\varphi^{(k)}(t) = \begin{cases} \displaystyle\int_0^t \frac{d^k}{dt^k} g(t,s)\psi(s)\, ds, & k \leq p-1 \\ \displaystyle\int_0^t \frac{d^p}{dt^p} g(t,s)\psi(s)\, ds + \psi(t), & k = p. \end{cases}$$

For $\varphi \in D = C_0^\infty(T)$ on $T = (0, \infty)$ with $\psi = \mathcal{L}\varphi$, it shows that the corresponding space $W = X^*$ for $X = \mathcal{L}^* L_2(T)$ is similar to the Sobolev space $\overset{\circ}{W}{}_2^p(T)$, namely, the norm

$$\|\varphi\|_W = \|\mathcal{L}\varphi\|_{L_2}$$

is such that

$$c \sum_{k=0}^{p} \int_{T_{\text{loc}}} |\varphi^{(k)}(t)|^2 \, dt \le \|\varphi\|_W^2 \le C \sum_{k=0}^{p} \int_T |\varphi^{(k)}(t)|^2 \, dt$$

for any finite interval $T_{\text{loc}} \subseteq T$.

Hence, we have the W-type test functions space

$$X = \mathcal{L}^* L_2(T) = [D], \qquad D = \overset{\bullet}{C}{}_0^\infty(T),$$

which is similar to the Sobolev space $\overset{\circ}{W}{}_2^{-p}(T)$. In particular, on any finite interval $S = (t_0, t_1)$, $[S] \subseteq T$, we have the corresponding Sobolev spaces

$$W(S) = W_2^p(S), \qquad X(S) = W_2^{-p}(S). \quad \blacksquare$$

From the given above resolvent representation

$$\varphi^{(k)}(t) = (g_t^{(k)}, \mathcal{L}\varphi) = (\mathcal{L}^* g_t^{(k)}, \varphi), \qquad \varphi \in D,$$

for $k = 0, \ldots, p-1$ with

$$g_t^{(k)}(s) = \begin{cases} \dfrac{d^k}{dt^k} g(t,s), & 0 \le s \le t, \\ 0, & s \ge t, \end{cases} \in L_2(T),$$

we see that our test functions space $X = \mathcal{L}^* L_2(T)$ contains delta-functions and their derivatives

$$\delta_t^{(k)} = \mathcal{L}^* g_t^{(k)}, \qquad k = 0, \ldots, p-1.$$

Employing a complete system δ_t, $t \in T$, in X we can treat any function $u \in \mathbf{W}$ as

$$u = u(t) \equiv (\delta_t, u), \qquad t \in T,$$

with

$$(\varphi, u) = \int_T \varphi(t) u(t) \, dt, \qquad \varphi \in D.$$

The random function $u = u(t)$, $t \in T$, is meansquare differential with the meansquare continuous derivatives

$$\frac{d^k}{dt^k}u(t) = u^{(k)}(t) = (-1)^k(\delta_t^{(k)}, u), \qquad k = 0, \dots, p-1.$$

Here $\mathcal{L}^*: L_2(S) \to X(S)$ is the nondegenerate operator (see lemma associated with condition (2.2)).

The deterministic equation (2.1) for any $f \in L_2(S)$ on $S = (t_0, t_1)$ and zero initial conditions

$$u^{(k)}(t_0) = 0, \qquad k = 0, \dots, p-1,$$

has the unique solution $u \in W(S) = W_2^p(S)$ as

$$u(t) = \int_{t_0}^{t} g(t,s) f(s)\, ds = (g_t^-, f), \qquad t \in S,$$

where

$$g_t^- = \begin{cases} g(t,s), & t_0 \le s \le t, \\ 0, & s < t_0, \ s > t_1. \end{cases}$$

Hence, there is a unique solution $u = \xi \in \mathbf{W}(S) = \mathbf{W}_2^p(S)$ of the stochastic equation (2.1) for any stochastic source $f = (\varphi, f)$, $\varphi \in C_0^\infty(S)$, meansquare continuous with respect to $\|\varphi\|_{L_2}$, and any stochastic initial conditions

$$u^{(k)}(t_0) = \xi_k, \qquad k = 0, \dots, p-1.$$

This solution $u = \xi \in \mathbf{W}(S)$ can be described by the general formula (2.16) which in our particular case gives

$$u(t) = (g_t^-, f) + \sum_{k=0}^{p-1} \xi_k u_k(t)$$

with a fundamental system $u_k(t)$, $k = 0, \dots, p-1$, of the equation $\mathcal{L}u = 0$ classical solutions,

$$u_k^{(j)}(t_0) = \begin{cases} 1, & j = k, \\ 0, & j \ne k. \end{cases}$$

To explain, we apply the resolvent decomposition

$$g(t,s) = \sum_{k=0}^{p-1} g_{t_0}^{(k)}(s) u_k(t), \qquad 0 \le s \le t_0,$$

which gives us

$$g_t = g_t^- + g_t^+, \qquad g_t^+ = \sum_{k=0}^{p-1} g_{t_0}^{(k)} u_k(t)$$

and

$$\delta_t = \mathcal{L}^* g_t = \mathcal{L}^* g_t^- + \sum_{k=0}^{p-1} \delta_{t_0}^{(k)} u_k(t).$$

It is worthy mentioning, that in a case of the stochastic source represented by a *stochastic Ito integral*

$$f = (\varphi, f) = \int \varphi(t)\, d\eta(t), \qquad \varphi \in C_0^\infty(S),$$

this solution $u = \xi \in \mathbf{W}(S)$ represents a random process $\xi = \xi(t)$, $t_0 \le t \le t_1$, described by *stochastic Ito equations*

$$\begin{aligned} d\xi(t) &= \xi^{(1)}(t)\, dt, \\ &\cdots\cdots\cdots\cdots \\ d\xi^{(p-2)}(t) &= \xi^{(p-1)}(t)\, dt, \\ d\xi^{(p-1)}(t) &= -\sum_{k=0}^{p-1} a_k \xi^{(k)}(t)\, dt + d\eta(t) \end{aligned}$$

with the initial conditions

$$\xi^{(k)}(t_0) = \xi_k, \qquad k = 0, \ldots, p-1. \quad \blacksquare$$

In what was shown, one can see that the initial conditions are of the type (2.17), (2.18) with $X^+(\Gamma)$ arising from outside $S = (t_0, t_1)$ by means of the boundary test functions

$$\delta_{t_0}^{(k)} = \mathcal{L}^* g_{t_0}^{(k)}, \qquad g_{t_0}^{(k)} \in L_2(0, t_0), \qquad k = 0, \ldots, p-1,$$

— cf. (2.9), etc. The p-dimensional boundary subspace $X^-(\Gamma)$ is generated by its basic elements

$$x_j^- = \mathcal{L}^* g_j^- = \sum_{k=0}^{p-1} [x_{jk}^-(t_0)\delta_{t_0}^{(k)} + x_{jk}^-(t_1)\delta_{t_1}^{(k)}], \qquad j = 0, \ldots, p-1,$$

which correspond to a fundamental system $g_j^- \in L_2(S)$, $j = 0, \ldots, p-1$, of the equation $\mathcal{L}^* g = 0$ classical solutions on $S = (t_0, t_1)$, $g_j^- = 0$ outside S; recall, the elements x_j^- serve the very equation (2.1) giving

$$(x_j^-, u) = \sum_{k=0}^{p-1} [x_{jk}^-(t_0) u^{(k)}(t_0) + x_{jk}^-(t_1) u^{(k)}(t_1)]$$

$$= (g_j^-, f), \qquad j = 0, \dots, p-1.$$

The whole boundary space $X(\Gamma) \subseteq X(S) = W_2^{-p}(S)$ is generated by its basic elements

$$\delta_t^{(k)}, \qquad t \in \Gamma = \{t_0, t_1\}, \qquad k = 0, \dots, p-1,$$

cf. (4.9), Chapter I, and any $X^+(\Gamma) \subseteq X(\Gamma)$ with its basic elements

$$x_j = \sum_{k=0}^{p-1} [x_{jk}(t_0)\delta_{t_0}^{(k)} + x_{jk}(t_1)\delta_{t_1}^{(k)}], \qquad j = 0, \dots, p-1,$$

linearly independent on x_j^-, $j = 0, \dots, p-1$, can serve for boundary conditions of the type (2.17), (2.18). Namely, we can set these boundary conditions as

$$\sum_{k=0}^{p-1} [x_{jk}(t_0)u^{(k)}(t_0) + x_{jk}(t_1)u^{(k)}(t_1)] = \xi_j, \qquad j = 0, \dots, p-1,$$

with arbitrary given ξ_j, $j = 0, \dots, p-1$. ■

We conclude here with a note, that the corresponding equation

$$\mathcal{P}u = f$$

on the interval $S = (t_0, t_1)$ with $\mathcal{P} = \mathcal{L}^*\mathcal{L}$ has a unique solution $u \in \mathbf{W}(S)$ for any stochastic source $f = (\varphi, f)$, $\varphi \in C_0^\infty(S)$, mean-square continuous with respect to the norm $\|\varphi\|_W$, and arbitrary given boundary conditions

$$u^{(k)}(t_0) = \xi_{0,k}, \qquad u^{(k)}(t_1) = \xi_{1,k}, \qquad k = 0, \dots, p-1,$$

for the boundary values

$$u^{(k)}(t) = (\delta_t^{(k)}, u), \qquad t \in \Gamma = \{t_0, t_1\}, \qquad k = 0, \dots, p-1,$$

determined by means of the boundary test functions

$$\delta_t^{(k)}, \qquad t \in \Gamma = \{t_0, t_1\}, \qquad k = 0, \dots, p-1.$$

This also concerns a general symmetric positive differential operator $\mathcal{P}$,

$$(\varphi, \mathcal{P}\bar{\varphi})^{1/2} \asymp \|\varphi\|_p, \qquad \varphi \in C_0^\infty(T_{\text{loc}}), \qquad T_{\text{loc}} \supseteq [S].$$

II.3. Boundary conditions for some random fields models

3.1. Stochastic string equation

Let us consider the differential equation

$$\frac{\partial^2 u}{\partial r^2} - \frac{\partial^2 u}{\partial s^2} = f \tag{3.1}$$

in the framework of our general scheme (2.1), (2.1)′ with the *operator*

$$\mathcal{L} = \frac{\partial^2}{\partial r^2} - \frac{\partial^2}{\partial s^2}$$

in $F = L_2(T)$, where the plane region $T \subseteq R^2$ of the time–space variable $t = (s, r)$ is taken as $T = \{(s, r) : |s| < r,\ r > 0\}$. This case is of special interest, since for a stochastic source $f = \eta$ being the Gaussian white noise we obtain the well-known Brownian Sheet $\xi = \xi(t)$, $t \in T$, as a unique solution $u = \xi \in \mathbf{W}$ of equation (3.1) — see (1.12), (1.12)′.

Let us recall that in other coordinates (t_1, t_2) which arise as a result of rotation of the R^2-plain by 45°, equation (3.1) has the form

$$\frac{\partial^2}{\partial t_1 \partial t_2} u = f \tag{3.1$'$}$$

with the differential operator

$$\mathcal{L} = \frac{\partial^2}{\partial t_1 \partial t_2}.$$

Dealing with equations (3.1) and (3.1)′ for $u \in \mathbf{W}(S)$ in the region $S \subseteq T$ we encounter the problem of finding boundary conditions in terms of the *boundary test functions* x, $\operatorname{supp} x \subseteq \Gamma$, with supports on Γ which can be set on the boundary $\Gamma = \partial S$. As we know, *all* these boundary test functions are $x = \mathcal{L}^* g$ with $g \in L_2(T)$

$$\mathcal{L}^* g = 0$$

in the region $T \setminus \Gamma$ outside Γ — see (2.9).

In particular, we have the delta-functions

$$x = \delta_s, \qquad s \in \Gamma, \tag{3.2}$$

at all boundary points $s \in \Gamma$ playing the role of the boundary test functions with the corresponding boundary values

$$(x, u) = (\delta_s, u) = u(s), \qquad s \in \Gamma. \tag{3.2$'$}$$

We show that in general, apart from the delta-functions (3.2), there are boundary test functions of the type

$$x = (\varphi, x) = \int_\Gamma x(s) \frac{\partial \varphi}{\partial l}\, ds, \qquad \varphi \in D = C_0^\infty(T), \tag{3.3}$$

which determine a *generalized trace* of the generalized nontangent derivative $\partial u/\partial l$ on the boundary Γ as

$$(x, u) = \int_\Gamma x(s) \frac{\partial u}{\partial l}\, ds \tag{3.3$'$}$$

according to the well-defined left-hand side with a variety of weight-functions $x(s)$, $s \in \Gamma$.

Let us consider a boundary part formed with

$$s = (s_1, s_2)\colon\ a_1 \leq s_1 \leq b_1, \quad s_2 = \gamma(s_1),$$

and take $g \in L_2(T)$ as $g = g(t_1)$ inside the region $a_1 < t_1 < b_1$, $t_2 < \gamma(t_1)$ and $g = 0$ outside; then we get the boundary test functions $x = \mathcal{L}^* g$ of the type

$$x = (\varphi, x) = (\mathcal{L}^* \varphi, g) = \int_{a_1}^{b_1} \Big[\int_0^{\gamma(t_1)} \frac{\partial^2}{\partial t_1 \partial t_2} \varphi(t_1, t_2)\, dt_2 \Big] g(t_1)\, dt_1 =$$

$$= \int_{a_1}^{b_1} \Big[\frac{\partial}{\partial t_1} \varphi(t_1, \gamma(t_2)) \Big] g(t_1)\, dt_1 = \int_\Gamma x(s) \frac{\partial}{\partial t_1} \varphi(s)\, ds, \qquad \varphi \in D,$$

with the corresponding weight-functions $x(s)$, $s \in \Gamma$. We can similarly proceed for a boundary part with

$$s = (s_1, s_2)\colon\ s_1 = \gamma(s_2), \quad a_2 \leq s_2 \leq b_2,$$

and get the boundary test functions

$$x = (\varphi, x) = \int_\Gamma x(s) \frac{\partial}{\partial t_2} \varphi(s)\, ds, \qquad \varphi \in D,$$

with the corresponding weight-functions $x(s)$, $s \in \Gamma$. On any part of the boundary Γ, where we can proceed in both ways, we get boundary test functions of the type (3.3) in terms of the sum

$$\int_\Gamma x(s) \cos\alpha_1 \frac{\partial}{\partial t_1}\varphi(t)\, ds + \int_\Gamma x(s) \cos\alpha_2 \frac{\partial}{\partial t_2}\varphi(s)\, ds$$

with the derivative

$$\frac{\partial}{\partial l} = \cos\alpha_1 \frac{\partial}{\partial t_1} + \cos\alpha_2 \frac{\partial}{\partial t_2}$$

along a direction $l = (\cos\alpha_1, \cos\alpha_2)$.

A different case arises for a boundary part with

$$s = (s_1, s_2)\colon\ s_1 = a_1, \quad a_2 \le s_2 \le b_2;$$

here we cannot get the derivative

$$\frac{\partial}{\partial l} = \frac{\partial}{\partial t_1}$$

at all, and the derivative

$$\frac{\partial}{\partial l} = \frac{\partial}{\partial t_2}$$

does not bring any complement to the collection of the delta-functions (3.2). A similar case arises for a boundary part with

$$s = (s_1, s_2)\colon\ a_1 \le s_1 \le b_1, \quad s_2 = a_2.$$

The test functions of the type (3.3) determine on the boundary the corresponding generalized derivative $\partial/\partial l$ for all functions $u \in \mathbf{W}$. In particular, for all deterministic functions $u \in W$ with the representation

$$u(t) = \int_0^{t_1}\int_0^{t_1} f\, dt_1\, dt_2, \qquad f \in L_2(T), \tag{3.4}$$

we can determine the trace $\partial/\partial l$ on the boundary Γ as

$$\frac{\partial u}{\partial l}(s) = \cos\alpha_1 \int_0^{s_2} f(s_1, t_2)\, dt_2 + \cos\alpha_2 \int_0^{s_1} f(t_1, s_2)\, dt_1, \qquad s \in \Gamma.$$

And it is worth mentioning again that random functions $u = \xi \in$ **W** are quite different from the deterministic ones. For example, the Browninan Sheet $\xi = \xi(t)$, $t \in T$, is no better than with

$$\|\xi(t) - \xi(s)\|^2 = \mathrm{E}|\xi(t) - \xi(s)|^2$$
$$= c_1|t_1 - s_1| + c_2|t_2 - s_2| - |t_1 - s_1| \times |t_2 - s_2|,$$
$$c_1 = \max(s_2, t_2), \qquad c_2 = \max(s_1, t_1),$$

for all $s, t \in T$. ■

Let us show that *a system of boundary test functions of the type* (3.2), (3.3) *is complete in the boundary test functions space* $X(\Gamma)$.

In our scheme with the operator $\mathcal{L} = \partial^2/\partial t_1 \partial t_2$ in the space $F = L_2(T)$ and the nondegenerate (unitary) operator $\mathcal{L}^* = \partial^2/\partial t_1 \partial t_2$ on $L_2(T)$, from the orthogonal decomposition

$$L_2(T) = L_2(S_-) + L_2(S_+)$$

for $S_- = S$ and a complementary region $S_+ = T \setminus [S]$, we obtain

$$X(\Gamma) = X^-(\Gamma) + X^+(\Gamma)$$

with the orthogonal components

$$X^-(\Gamma) = \mathcal{L}^* L_2(S) \cap X(\Gamma), \qquad X^+(\Gamma) = \mathcal{L}^* L_2(S_+) \cap X(\Gamma).$$

And it will be enough to show that the deterministic function $u \in W$ is uniquely determined in each region S_-, S_+ by equation (3.1)′ and boundary values (3.2)′, (3.3)′, with boundary test functions (3.2), (3.3) from the corresponding $X^+(\Gamma)$, $X^-(\Gamma)$. Here one can take into account the fact that $X^-(\Gamma)$ plays the same role of the test functions subspace with respect to S_+ as $X^+(\Gamma)$ does with respect to the very region $S = S_-$ (see (2.5)–(2.9)).

Let us show that for the region $S = (a_1, b_2) \times (a_2, b_2)$ of the rectangular type (see Fig. 3) the *complete* system is formed by only test functions of the type (3.2). Equation (3.1)′ with the boundary values $u(s)$ on the boundary parts $s_1 = a_1$, $a_2 < s_2 < b_2$, and $a_1 \le s \le b_1$, $s_2 = a_2$, completely determine $u \in W(S)$ in the region S, bacause, according to the representation (3.4) with $f = \mathcal{L}u \in L_2(T)$, we have

$$u(t) = u(a_1, t_2) + u(t_1, a_2) - u(a_1, a_2) + \int_{a_1}^{t_1}\int_{a_2}^{t_2} f(t)\, dt, \qquad t \in S.$$

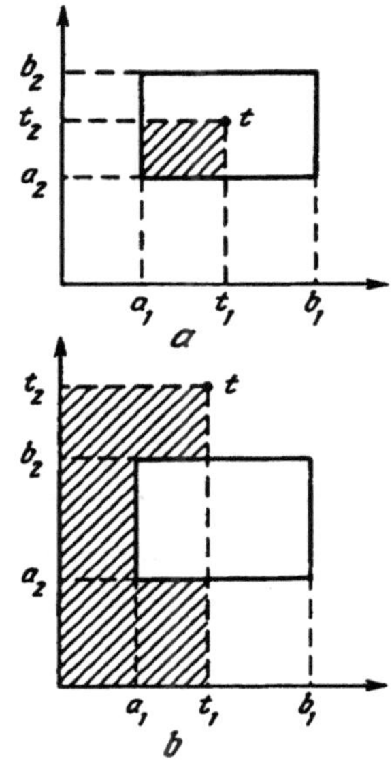

Fig. 3

Similarly we can proceed with equation (3.1)′ in the complementary region S_+ with the boundary values $u(s)$ on the boundary parts $a_1 \leq s_1 \leq b_1$, $s_2 = b_2$, and $s_1 = b_1$, $a_2 \leq s_2 \leq b_2$, and get $u \in W(S_+)$ in S_+. For example, according to Fig. 3 we have

$$u(t) = u(t_1, b_2) - u(a_1, b_2) - u(t_1, a_2) + u(a_1, a_2) +$$
$$+ \int_0^{t_1}\int_0^{t_2} f(t)\, dt - \int_{a_1}^{t_1}\int_{a_2}^{b_2} f(t)\, dt.$$

Let us now take the region $S \subseteq T$ of the general type (see Fig. 4) with the trace of $u \in W$ and its *nontangent* derivative $\partial u/\partial l$ on a

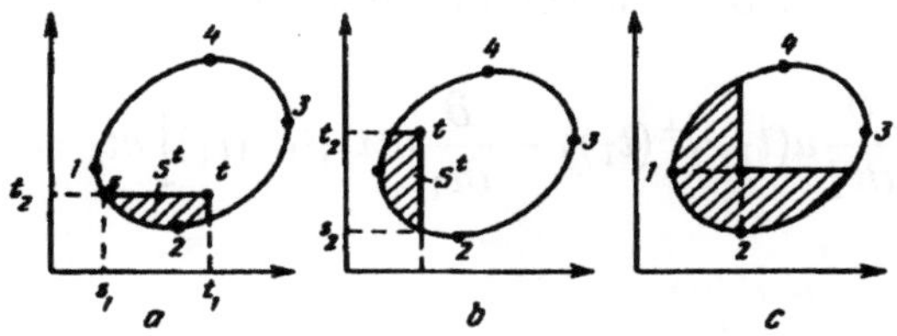

Fig. 4

boundary part (1-2) plus the trace $\partial u/\partial t_1$ on (2-3) and $\partial u/\partial t_2$ on (1-4) all given in (3.2)′, (3.3)′. It is easy to see that these boundary data and equation (3.1)′ in S determine $u \in W(S)$ in the region S. For example, with the representation (3.4) we have

$$u(t) = \int_{s_1}^{t_1} \frac{\partial}{\partial t_1} u(t_1, \gamma(t_1))\, dt_1 + u(s) + \iint_{S^t} f(t)\, dt$$

according to Fig. 4a and

$$u(t) = \int_{s_2}^{t_2} \frac{\partial}{\partial t_2} u(\gamma(t_2), t_2)\, dt_2 + u(s) + \iint_{S^t} f(t)\, dt$$

according to Fig. 4b, where

$$\int_{s_1}^{t_1} \frac{\partial}{\partial t_1} u(t_1, \gamma(t_1))\, dt_1 = \int_{s_1}^{t_1} \int_{0}^{\gamma(t_1)} f\, dt_2\, dt_1,$$

$$\int_{s_2}^{t_2} \frac{\partial}{\partial t_2} u(\gamma(t_2), t_2)\, dt_2 = \int_{s_2}^{t_2} \int_{0}^{\gamma(t_2)} f\, dt_1\, dt_2$$

on the boundary parts represented correspondingly as $t_2 = \gamma(t_1)$ or $t_1 = \gamma(t_2)$. In this way we determine $u \in W(S)$ in the region shown in Fig. 4c and we can proceed further in a similar way to determine $u \in W$ in S. Considering equation (3.1)′ in S_+, we can set the trace of the derivatives $\partial u/\partial t_1$, $\partial u/\partial t_2$ in the form

$$\int_{a_1}^{t_1}\Big[\frac{\partial}{\partial t_1}u(t_1,\gamma^+(t_1)) - \frac{\partial}{\partial t_1}u(t_1,\gamma^-(t_1)\Big]\,dt_1 =$$

$$= \int_{a_1}^{t_1}\int_{\gamma^-(t_1)}^{\gamma^+(t_1)} f\,dt_1\,dt_2, \qquad a_1 \le t_1 \le b_1,$$

on the boundary part (1-4-3) of $t_2 = \gamma^+(t_1)$ and on (1-2-3) of $t_2 = \gamma^-(t_1)$ plus

$$\int_{a_2}^{t_2}\Big[\frac{\partial}{\partial t_2}u(\gamma^+(t_2),t_2) - \frac{\partial}{\partial t_2}u(\gamma^-(t_2),t_2)\Big]\,dt_2 =$$

$$= \int_{a_2}^{t_2}\int_{\gamma^-(t_2)}^{\gamma^+(t_2)} f\,dt_1\,dt_2, \qquad a_2 \le t_2 \le b_2,$$

on the boundary part (2-3-4) of $t_1 = \gamma^+(t_2)$ and on (2-1-4) of $t_1 = \gamma^-(t_2)$. As easy to see, we obtain

$$u(t) = \int_{a_1}^{t_1}\Big[\frac{\partial}{\partial t_1}u(t_1,\gamma^+(t_1)) - \frac{\partial}{\partial t_1}u(t_1,\gamma^-(t_1)\Big]\,dt_1 + \iint_{S_+^t} f\,dt_1\,dt_2$$

— see Fig. 5a

$$u(t) = \int_{a_2}^{t_2}\Big[\frac{\partial}{\partial t_2}u(\gamma^+(t_2),t_2) - \frac{\partial}{\partial t_2}u(\gamma^-(t_2),t_2)\Big]\,dt_2 + \iint_{S_+^t} f\,dt_1\,dt_2,$$

— see Fig 5b, and proceeding further in a similar way, we can determine $u \in W(S_+)$ in the region S_+. ■

Our approach to boundary values of the generalized random functions $u \in \mathbf{W}$ makes it possible to consider boundary problems for the stochastic string equation (3.1).

For example, in the region S of the rectangular type (see Fig. 6) formed by characteristics of the equation (3.1) we can set the so-called *Goursat problem* with an arbitrary trace of $u \in \mathbf{W}(S)$

$$u = u^+$$

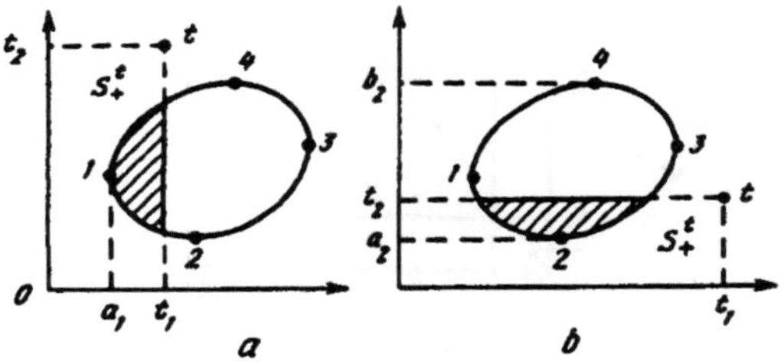

Fig. 5

on the boundary part

$$\{(a_1, b_1) \times a_2\} \cup \{a_1 \times (a_2, b_2)\}.$$

In our framework, this is the boundary problem with boundary conditions

$$(x, u) = (x, u^+), \qquad x \in X^+(\Gamma),$$

of the type $X^+(\Gamma) = X_0^+(\Gamma)$ (see (2.6) and (2.10)), and there is a unique solution $u \in \mathbf{W}(S)$ for any *stochastic* source $f = (\varphi, f)$, $\varphi \in C_0^\infty(S)$, meansquare continuous with respct $\|\varphi\|_{L_2}$, and an arbitrary *stochastic* sample $u^+ \in \mathbf{W}(S)$.

In the region S of the general type (see the Fig. 7) it was actually considered another boundary problem of the type $X^+(\Gamma) = X_0^+(\Gamma)$ with an arbitrary given trace

$$u = u^+, \qquad \frac{\partial u}{\partial l} = \frac{\partial u^+}{\partial l} \tag{3.5}$$

of $u \in \mathbf{W}(S)$ and its generalized nontangent derivative $\partial u/\partial l$ on the boundary part (1-2) plus

$$\frac{\partial u}{\partial l} = \frac{\partial u^+}{\partial l} \qquad (3.5)'$$

with $\partial/\partial l = \partial u/\partial t_1$ on the boundary part (2-3) and $\partial/\partial l = \partial/\partial t_2$ on (1-4). This is some version of the so-called *second mixed boundary problem* which is mostly considered in the cylinder-like region S (see Fig. 8).

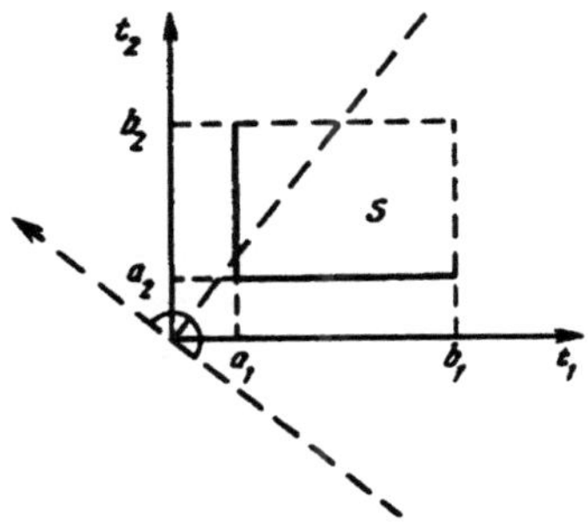

Fig. 6

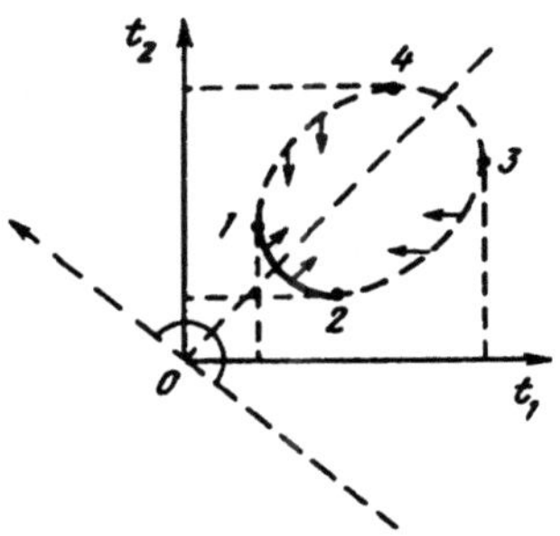

Fig. 7

The so-called *first mixed boundary problem* consists in finding a solution of the equation (3.1) with the boundary conditions (3.5) on the boundary part (1-2) plus

$$u = u^+$$

on the boundary part (2-3), (1-4). The general *second mixed bound-*

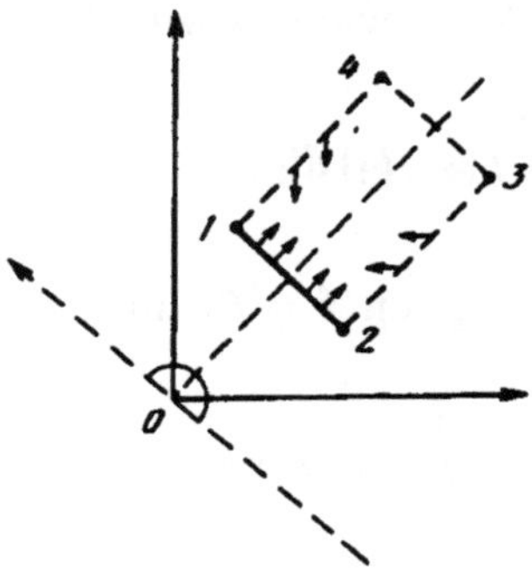

Fig. 8

ary problem is of the type (3.5), (3.5)′ with arbitrary $\partial/\partial l$ in (3.5)′, $\partial/\partial l \neq \partial/\partial t_1$ on (1-4) and $\partial/\partial l \neq \partial/\partial t_2$ on (2-3). There is also the *third mixed boundary problem* with the boundary conditions (3.5) on (1-2) plus a linear combination

$$c_1 u + c_2 \frac{\partial u}{\partial l} = c_1 u^+ + c_2 \frac{\partial u^+}{\partial l}$$

on (2-3), (1-4).

In our scheme, all these boundary problems are of the type $X^+(\Gamma)$ with the known direct sum representation for the test functions space $X(S) = [C_0^\infty(S)]$ in the region S,

$$X(S) = \mathcal{L}^* L_2(S) + X^+(\Gamma).$$

Hence, for all these boundary problems the following result holds true.

THEOREM. *There is a unique solution $u \in \mathbf{W}(S)$ for any stochastic source in the equation, $f = (\varphi, f)$, $\varphi \in C_0^\infty(S)$, meansquare continuous with respect to $\|\varphi\|_{L_2}$ and arbitrary stochastic sample $u^+ \in \mathbf{W}(S)$ in the boundary conditions.*

To explain, we turn on equation (3.1)′ and apply Fig. 9. The deterministic function $u \in \mathbf{W}(S)$ plus its nontangent derivative $\partial u/\partial l$

on the boundary part (1-2) provide us with the derivative

$$\frac{\partial u}{\partial t_1}(s) = \int_0^{s_2} f(s_1, t_2)\, dt_2, \qquad s = (s_1, s_2).$$

Apart from that, from the given $f = \mathcal{L}u$ in S we can get

$$\int_{s_2}^{s_2^+} f(s_1, t_2)\, dt_2$$

and

$$\frac{\partial u}{\partial t_1}(s^+) = \int_0^{s_2^+} f(s_1, t_2)\, dt_2, \qquad s^+ = (s_1, s_2^+),$$

on the boundary part of (1-4) just above (1-2). In the case of the second boundary problem, say, with given

$$\frac{\partial u}{\partial l} = c_1 \frac{\partial u}{\partial t_1} + c_2 \frac{\partial u}{\partial t_2}$$

($c_2 \neq 0$) on (1-4) we also get

$$\frac{\partial u}{\partial t_2} = \frac{1}{c_2}\Big(\frac{\partial u}{\partial l} - c_1 \frac{\partial u}{\partial t_1}\Big).$$

In a similar way, we can find

$$\frac{\partial u}{\partial t_1} = \frac{1}{c_1}\Big(\frac{\partial u}{\partial l} - c_2 \frac{\partial u}{\partial t_2}\Big)$$

on the boundary part (2-3) right to (1-2), and so on, which finally allows us to reduce the general boundary problem to the boundary problem (3.5), (3.5)′.

3.2. Stochastic Laplace equation

Let us consider a differential equation

$$\Delta u = f \tag{3.6}$$

applying our scheme for the Laplace operator $\mathcal{L} = \Delta$,

$$\Delta = \sum_k \frac{\partial^2}{\partial_k^2},$$

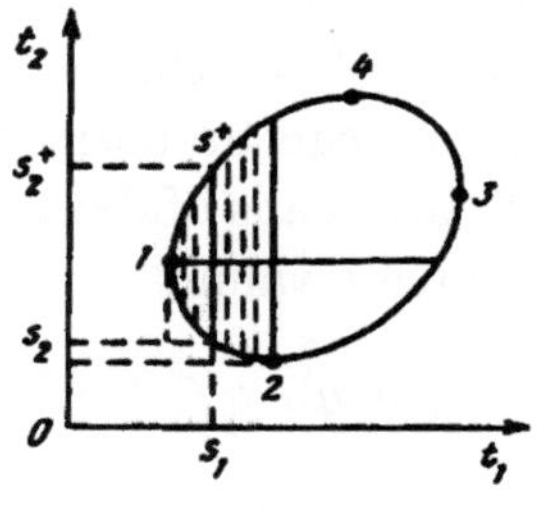

Fig. 9

in the space $F = L_2(T)$. Recall, in a case of $T = R^3 \setminus \{0\}$, with Gaussian white noise $f = \eta$ it gives, as a unique solution $u = \xi \in \mathbf{W}$, a random field $\xi = \xi(t)$, $t \in T$, representing Levy Brownian motion — see (1.13), etc. This is a particular example of random fields which can be described by stochastic elliptic differential equations, and there one can be interested in setting appropriate boundary conditions. (We mentioned already about elliptic boundary problems, related to general boundary conditions of the type (2.11).)[3] Here we would like to stress on a very particular boundary problem for the equation (3.6) in a region S, $[S] \subseteq T$, with the stochastic source $f = \eta$ represented by a generalized random field $\eta = (\varphi, \eta)$, $\varphi \in C_0^\infty(S)$, with independent values, meansquare continuous with respect to the norm $\|\varphi\|_{L_2}$. Note, that in the scheme considered for the bounded region S we have the test functions space $X(S) = W_2^{-2}(S)$, since in any bounded region $T_{\mathrm{loc}} \subseteq T$, $[S] \subseteq T_{\mathrm{loc}}$, we have, correspondingly, $W = [C_0^\infty(T_{\mathrm{loc}})]$ with the norm

$$\|\varphi\|_W = \|\Delta\varphi\|_{L_2} \asymp \|\varphi\|_2, \qquad \varphi \in C_0^\infty(T_{\mathrm{loc}}),$$

as the Sobolev space $W = \overset{\circ}{W}{}_2^2(T_{\mathrm{loc}})$ and $W(S) = X(S)^* \subseteq W$, $X(S) \subseteq X = W^*$ as the Sobolev spaces

$$W(S) = W_2^2(S), \qquad X(S) = W_2^{-2}(S).$$

[3] Cf., for example, G. Benfatto, G. Gallavotti, and F. Nicolo: Elliptic equations and Gaussian processes. J. Funct. Anal., 36 (1980), 343–400.

Namely, considering the generalized random field described as a unique solution $u = \xi \in \mathbf{W}$ of the equation (3.6) in T with a stochastic source $f = \eta$, one can be interested in a forecast of ξ in the subregion S, $[S] \subseteq T$, by means of data with respect to η outside S, that is, by means of outside values

$$\mathcal{B} = \{(g, \eta),\ g \in L_2(S_+)\}$$

in a complementary region $S_+ = T \setminus [S]$. As known, the best forecast is to be given as the corresponding conditional expectation

$$\hat{\xi} = \mathrm{E}(\xi/\mathcal{B}),$$

$$(\varphi, \hat{\xi}) = \mathrm{E}[(\varphi, \xi)/\mathcal{B}], \qquad \varphi \in C_0^\infty(S).$$

Here we have $u = \xi \in \mathbf{W}(S)$ in the stochastic Sobolev space $\mathbf{W}(S) = \mathbf{W}_2^2(S)$. For the stochastic source $f = \eta$ with zero mean and independent values in S and S_+, taking the conditional expectation in the very equation (3.6), we obtain, for another $u = \hat{\xi} \in \mathbf{W}(S)$ in the stochastic Sobolev space $\mathbf{W}(S) = \mathbf{W}_2^2(S)$, the *stochastic Laplace equation*

$$\Delta u = 0$$

in the region S. Can we transfer the given data $\mathcal{B}$ in some kind of boundary conditions on $\Gamma = \partial S$ which allow to determine the best forecast as the unique solution $u = \hat{\xi} \in \mathbf{W}(S)$? The answer is "yes." Namely, we can apply boundary test functions $x = \mathcal{L}^* g$, taking $g \in L_2(T)$ such that $g = 0$ in S and g are harmonic functions in S_+ ($\mathcal{L}^* g = 0$ for $\mathcal{L}^* = \Delta$). These $x \in X(\Gamma)$ form a certain complement $X^+(\Gamma)$ to the test functions subspace $X^-(S) = \mathcal{L}^* L_2(S)$ serving the general differential equation (3.6) in the region S (see (2.8)). In particular, for the nondegenerate operator $\mathcal{L}^* = \Delta$ on $L_2(T)$, — see a case of Levy Brownian motion, we obtain

$$X^+(\Gamma) = X_0^+(\Gamma)$$

as the orthogonal complement to $X^-(S)$ in the test functions space $X(S)$. Hence, for arbitrary given boundary conditions

$$(x, u) = (x, u^+), \qquad x \in X^+(\Gamma),$$

on the boundary $\Gamma = \partial S$, the equation (3.6) has a unique solution $u \in \mathbf{W}(S)$.

Note, that we can identify our boundary test functions $x = \Delta g \in X^+(\Gamma)$ by means of a proper limit Green formula as Schwartz distributions

$$x = (x, \varphi) = \int_\Gamma (\partial g\, \varphi - g\, \partial\varphi)\, ds, \qquad \varphi \in C_0^\infty(T),$$

with the boundary trace of φ and their normal derivatives $\partial\varphi$ on Γ. Employing this notations for x and taking the known boundary trace

$$u = u^{(0)}, \qquad \partial u = u^{(1)}$$

for $u \in \mathbf{W}(S) = W_2^2(S)$, we can describe the boundary conditions as

$$\int_\Gamma (\partial g\, u - g\, \partial u)\, ds = \int_\Gamma (\partial g\, u^+ - g\, \partial u^+)\, ds,$$

where we can take any

$$u^{(0)} \in \mathbf{W}_2^{3/2}(\Gamma), \qquad u^{(1)} \in \mathbf{W}_2^{1/2}(\Gamma)$$

from the corresponding stochastic Sobolev spaces on the boundary $\Gamma = \partial S$. The most direct way to set these boundary conditions would be to employ the very boundary values

$$(x, u) = (g, f), \qquad x = \mathcal{L}^* g, \tag{3.7}$$

which result from testing f in the complementary region $S_+ = T \setminus [S]$ by the harmonic functions $g \in L_2(S_+)$.

Thus, considering our forecast problem in a case of the stochastic source $f = \eta$ being with independent values, we can apply the following.

THEOREM. *The best forecast*

$$\hat{\xi} = \mathrm{E}[\xi/(g, \eta),\, g \in L_2(S_+)]$$

can be obtained as the unique solution $u = \hat{\xi} \in \mathbf{W}(S)$ in the stochastic Sobolev space $\mathbf{W}(S) = \mathbf{W}_2^2(S)$ of the Laplace equation (3.6) *with $f = 0$ in the region S and the boundary conditions* (3.7) *which result from testing the stochastic source $f = \eta$ outside S by the harmonic functions g in the complementary region S_+.*

II.4. Boundary Conditions for Stochastic Evolution Equations

4.1. Preliminaries

In this section we are going to apply our general approach to boundary conditions with respect to the *stochastic evolution equation*

$$d\xi_t = A\xi_t\,dt + \eta_{dt} \tag{4.1}$$

where A is a *symmetric* differential operator

$$A = \sum_{|k|\le p} a_k \partial^k = \sum_{|k|\le p} (-1)^{|k|}\partial^k a_k = A^* \le 0$$

in the region $G \subseteq R^{d-1}$, a stochastic source η is of the white noise type, say, and

$$\xi = \xi_t, \qquad t \in I = (t_0, t_1),$$

represents the generalized random field $\xi \in \mathbf{W}(G\times I) = \mathbf{W}_2^{(p,1)}(G \times I)$ in the known stochastic Sobolev space on a cylinder $G \times I \subseteq R^d$ with its *generalized trace*

$$\xi_t = (\varphi, \xi_t), \qquad \varphi \in C_0^\infty(G).$$

Recall, equation (4.1) means that

$$(\varphi, \xi_b) - (\varphi, \xi_a) = \int_a^b (A\varphi, \xi_t)\,dt + \int_a^b (\varphi, \eta_{dt}),$$

$$\varphi \in C_0^\infty(G), \qquad t_0 < a < b < t_1,$$

and, as was actually shown, it is equivalent to the generalized differential equation

$$\frac{\partial u}{\partial t} = Au + f \tag{4.1$'$}$$

in the cylinder $G \times I$ with $u = \xi$ as

$$u = (\varphi, u) = \int_I \varphi_t \xi_t\,dt$$

and f as

$$f = (\varphi, f) = \int_I \varphi_t\,d\eta_t, \qquad \varphi \in C_0^\infty(G \times I),$$

where f is mean square continuous with respect to $\|\varphi\|_{L_2}$, $\varphi \in C_0^\infty(G\times I)$, and well-defined on the whole space $L_2(G\times I)$. Here, giving a priority to the equation (4.1), we employ a traditional notation for the time variable $t \in I$ — cf. (1.18), (1.18)′.

Having in mind an appropriate *self-adjoint* extension $A \leq 0$ of the symmetric operator $A\colon C_0^\infty(G) \to L_2(G)$ in the $L_2(G)$-space, one can be interested in a particular solution $\xi = \xi_t$, $t \in I$, of equations (4.1), (4.1)′ given by the corresponding *semigroup formula*

$$\xi_t = \int_{t_0}^{t} e^{A(t-s)}\,\eta_{ds} \tag{4.2}$$

as

$$(\varphi, \xi_t) = \int_{t_0}^{t} [e^{A(t-s)}\varphi]\,\eta_{ds} = (1_{(t_0,t)}(s)e^{A(t-s)}\varphi, f)$$

for all $\varphi \in C_0^\infty(G)$ and $t \in I$, where, for the time variable $s \in I$

$$1_{(t_0,t)}(s)e^{A(t-s)}\varphi \in L_2(G\times I).$$

Note that as a result of the formula (4.2) we find

$$\xi_t = \int_{t_0}^{t} A\xi_s\,ds + \int_{t_0}^{t} \eta_{ds}, \qquad t \in I,$$

represented in terms of the stochastic integrals (see Section 1.2, Chapter I) as

$$(\varphi, \xi_t) = \int_{t_0}^{t} (A\varphi, \xi_s)\,ds + \int_{t_0}^{t} \varphi\,\eta_{ds}, \qquad \varphi \in C_0^\infty(G),$$

for

$$\xi_t - \int_{t_0}^{t} \eta_{ds} = \int_{t_0}^{t} [e^{A(t-s)} - 1]\,\eta_{ds} =$$

$$= \int_{t_0}^{t} \left[\int_{s}^{t} \frac{d}{dr} e^{A(r-s)}\,dr\right] \eta_{ds} = \int_{t_0}^{t} \left[\int_{t_0}^{r} \frac{d}{dr}\, e^{A(r-s)}\,\eta_{ds}\right] dr =$$

$$= \int_{t_0}^{t} \left[A \int_{t_0}^{r} e^{A(r-s)} \, \eta_{ds} \right] dr = \int_{t_0}^{t} A\xi_r \, dr$$

We assume that the differential operator A, considered in some region $G_0 \supseteq [G]$, is of a type

$$\|A\varphi\|_{L_2}^2 = (\varphi, A^*A\varphi) \asymp \|\varphi\|_p^2 = \sum_{|k|\le p} \|\partial^k \varphi\|_{L_2}^2, \qquad \varphi \in C_0^\infty(G_0).$$

This allows to apply our general scheme for the differential operator

$$\mathcal{L} = \frac{\partial}{\partial t} - A$$

in the L_2-space with the corresponding test functions space

$$X(S) = W_2^{-(p,1)}(S) = [C_0^\infty(S)]$$

in the cylinder $S = G \times I$ serving equations (4.1), (4.1)′. With the same self-adjoint extension $A \le 0$ as in the semigroup formula (4.2) we associate certain boundary conditions of the type (2.17), (2.18) on the cylinder boundary $\Gamma = \partial S$; this type of boundary conditions leads us to the direct sum representation

$$X(S) = \mathcal{L}^* L_2(S) + X^+(\Gamma)$$

with the corresponding boundary subspace $X^+(\Gamma)$ where the boundary conditions can be set as

$$(x, u) = (x, u^+), \qquad x \in X^+(\Gamma), \tag{4.3}$$

with any random sample $u^+ \in \mathbf{W}(S) = \mathbf{W}_2^{(p,1)}(S)$. Then the unique solution $u = \xi \in \mathbf{W}(S) = \mathbf{W}_2^{(p,1)}(S)$ in $S = G \times I$ can be described as

$$(\varphi, u) = (g, f) + (x, u^+), \qquad \varphi \in C_0^\infty(S),$$

with the test functions φ represented in the form

$$\varphi = \mathcal{L}^* g + x, \qquad g \in L_2(S), \qquad x \in X^+(\Gamma),$$

— cf. (2.16). The particular solution $u = \xi$ given by the semigroup formula (4.2) satisfies the *zero* boundary conditions (4.3) with the sample $u^+ = 0$. ■

Let a domain D_A of a self-adjoint extension

$$A\colon C_0^\infty(G) \subseteq D_A \to L_2(G),$$

be such that

$$D_A \subseteq W(G) = W_2^p(G),$$

where $W(G) = W_2^p(G)$ is the Sobolev space associated with the differential operator A in the space $F = L_2(G_0)$, $G_0 \supseteq [G]$. Having in mind this kind of self-adjoint operator $A \leq 0$, we can describe the corresponding type of boundary conditions (4.3) as follows.

The domain D_A of the closed operator A is complete with respect to

$$\|u\|^2_{L_2(G)} + \|Au\|^2_{L_2(G)} \leq C\|u\|^2_{W(G)}, \qquad u \in D_A,$$

so $D_A \subseteq W(G)$ is a (closed) subspace in $W(G)$,

$$\|u\|^2_{L_2(G)} + \|Au\|^2_{L_2(G)} \asymp \|u\|^2_{W(G)}. \tag{4.4}$$

The dual space $X(G) = W(G)^*$ is a closure

$$X(G) = [C_0^\infty(G)] = W_2^{-p}(G)$$

of all test functions $x \in C_0^\infty(G)$,

$$x = (u, x) \equiv (x, u), \qquad u \in W(G),$$

and $x \in X(G)$ can be identified as the certain Schwartz distributions in the region $G_0 \supseteq [G]$ with support

$$\operatorname{supp} x \subseteq [G]$$

in the closure of G. Here we recall that $X(G)$ can be regarded as the closure of $C_0^\infty(G)$ in the corresponding

$$X = W^* = \overset{\circ}{W}{}_2^{-p}(G_0)$$

formed by the Schwartz distributions

$$x = (u, x) \equiv (u, x), \qquad u \in C_0^\infty(G_0),$$

continuous with respect to

$$\|u\|_W = \|Au\|_{L_2} \asymp \|u\|_p$$

in $W = [C_0^\infty(G_0)] = \overset{\circ}{W}{}_2^p(G_0)$.

The subspace $D_A \subseteq W(G)$ can be described by means of its *annulator* $X^+(\partial G)$ in the dual space $X(G)$ which is formed by some $x \in X(G)$ with supports $\operatorname{supp} x \subseteq \partial G$ on the boundary ∂G, because

$$(u, x) \equiv (x, u) = 0, \qquad u \in D_A \supseteq C_0^\infty(G).$$

Thus, with the annulator $X^+(\partial G)$ to hand, we can describe the domain D_A as

$$D_A = \{u \in W(G)\colon\ (x, u) = 0,\ x \in X^+(\partial G)\}$$

in terms of the *zero* boundary conditions

$$(x, u) = 0, \qquad x \in X^+(\partial G). \tag{4.5}$$

Here, the boundary test functions $x \in X^+(\partial G)$ result distributions

$$(x, u) = \int_{\partial G} \sum_{|k| \le p-1} x_k \partial^k u(s)\, ds$$

of $u \in W_2^p(G)$ and its nontangent generalized derivatives $\partial^k u$, $|k| \le p-1$ on the boundary ∂G with the corresponding weight-functions $x_k(s)$, $s \in \partial G$ — see Sections 4.3, 5.3, Chapter I. ■

Now we can describe the space $X^+(\Gamma)$ related to the boundary conditions (4.3) on the boundary $\Gamma = \partial S$ of the cylinder $S = G \times I$, $I = (t_0, t_1)$, as a collection of boundary test functions of the type

$$x \times \delta, \qquad x \in X^+(\partial G), \qquad \delta \in C_0^\infty(I) \tag{4.6}$$

on the side-boundary $\partial G \times I$ and

$$\varphi \times \delta_{t_0}, \qquad \varphi \in C_0^\infty(G), \tag*{(4.6)$'$}$$

on the base-boundary $G \times \{t_0\}$ with the delta-function δ_{t_0} at the initial point $I = (t_0, t_1)$. The generalized test functions (4.6), (4.6)$'$ are well-defined in an appropriate cylinder

$$G_0 \times (a, b) = T \supset S = G \times (t_0, t_1),$$

where we can apply our known test functions space

$$X = \mathcal{L}^* L_2(T) = [C_0(T)] = \overset{\circ}{W}{}_2^{-(p,1)}(T)$$

of all Schwartz distributions $x = (u, x)$, $u \in C_0^\infty(T)$, which are continuous with respect to the norm $\|u\|_W$ of $u = u_t$, $t \in I$,

$$\|u\|_W^2 = \|\mathcal{L}u\|_{L_2(T)}^2 = \int_a^b \Big[\Big\|\frac{\partial}{\partial t}u_t\Big\|_{L_2(G_0)}^2 + \|Au_t\|_{L_2(G_0)}^2\Big]\,dt \asymp \|u\|_{(p,1)}^2,$$

in the corresponding space

$$W = \overset{\circ}{W}(T) = [C_0^\infty(T)] = \overset{\circ}{W}{}_2^{(p,1)}(T);$$

to explain, we expose

$$\|\mathcal{L}u\|_{L_2(T)}^2 = (u, \mathcal{P}\bar{u}), \qquad \mathcal{P} = \mathcal{L}^*\mathcal{L} = -\frac{\partial^2}{\partial t^2} + A^2$$

for the symmetric differential operator $A = A^*$.

The boundary test functions (4.6), (4.6)′ generate the boundary subspace $X^+(\Gamma)$ related to the boundary conditions (4.3) which can be roughly described by means of the corresponding weight-functions $x_k(s)$, $s \in \partial G$, for $u = u_t$, $t \in I$, as distributions

$$(x, u_t) = \int_{\partial G} \sum_{|k| \le p-1} x_k \partial^k u_t(s)\,ds =$$

$$= \int_{\partial G} \sum_{|k| \le p-1} x_k \partial^k u_t^+(s)\,ds = (x, u_t^+), \qquad t \in I, \tag{4.7}$$

on the side-boundary $\partial G \times I$, and

$$u_{t_0}(s) = u_{t_0}^+(s), \qquad s \in G, \tag{4.7$'$}$$

on the base-boundary $G \times \{t_0\}$ of the cylinder $S = G \times (t_0, t_1)$, — cf. (4.23), (4.23)′, Chapter I. ■

For a symmetric differential operator A in the region G there is a question on whether certain boundary conditions on the boundary ∂G determine the domain $D_A \subseteq W(G)$ of its self-adjoint extension A. To justify these conditions, we can proceed, for example, as follows. First, given the boundary conditions of the type (4.5) for $u \in W(G)$ on the boundary ∂G, we can verify (by integration by parts, say) that on D_A we have the symmetric operator $A \le 0$. Employing this symmetric operator $A: D_A \to L_2(G)$, in the case when some eigenfunctions $u_0 \in D_A$: $Au_0 = 0$ exist and form a closed subspace $D_A^0 \subseteq D_A$, we can check solvability of the equation

$$Au = f, \qquad u \in D_A \tag{4.8}$$

for any $f \in L_2(G)$, $f \perp D_A^0$; if it be so, then A is surely self-adjoint. Indeed, for any $h, g \in L_2(G)$ such that

$$(Av, h) = (v, g), \qquad v \in D_A$$

for all $v \perp D_A^0$, with the representation $g = u_0 + Au$ we obtain

$$(Av, h) = (v, Au) = (Av, u), \qquad (Av, h - u) = 0$$

which implies $h - u = v_0 \in D_A^0$, $h = u + v_0 \in D_A$.

For illustration, let us consider $A = d^2/dr^2$ on the interval $G = \{a < r < b\}$. With the boundary test functions

$$x = \delta_r, \delta_r', \qquad r \in \partial G = \{a, b\},$$

the boundary conditions (4.5) involve the corresponding boundary values

$$(x, u) = u(r), u'(r), \qquad r \in \partial G = \{a, b\}$$

and determine the symmetric extension A in a case of its domain described as

$$[u'(b)v(b) - u'(a)v(a)] - [u(b)v'(b) - u(a)v'(b)] = 0, \qquad u, v \in D_A.$$

To justify that A is self-adjoint, we are to take in account the corresponding boundary subspace $X^-(\partial G)$ we know about as genrated by $x = A^*g$, $g \in L_2(G_0)$ in $G_0 \supseteq [G]$ such that $A^*g = 0$ in G and $g = 0$ outside G; for $A = A^* = \partial^2/\partial r^2$ these boundary test functions are

$$x = \begin{cases} \delta_b' - \delta_a' & \text{for } g = 1, \ a < r < b, \\ (b\delta_b' - a\delta_a') - (\delta_b - \delta_a) & \text{for } g = r, \ a < r < b. \end{cases}$$

And with any boundary subspac $X^+(\partial G)$ in the boundary conditions (4.5) giving us the symmetric operator A, when this $X^+(\partial G)$ is in a complement to $X^-(\partial G)$, we have the equation (4.8) solvable for all $f \in L_2(G)$ giving the corresponding $u \in D_A$. Hence, in this case we have the self-adjoint operator A. For example, it holds true for the *Dirichlet type boundary conditions*

$$(\delta_r, u) = u(r) = 0, \qquad r \in \partial G = \{a, b\}.$$

A different case can be illustrated with the *Neuman type boundary conditions*

$$(\delta'_r, u) = u'(r) = 0, \qquad r \in \partial G = \{a, b\}$$

which, in particular, involve

$$u'(b) - u'(a) = (\delta'_b - \delta'_a, u) = 0$$

with

$$x = \delta'_b - \delta'_a \in X^-(\partial G).$$

Here, as we know, for solvability of the equation (4.8) the boundary conditions must be *consistent* with the corresponding $f \in L_2(g)$, i.e., in our case

$$(x, u) = u'(b) - u'(a) = \int_a^b f(r)\, dr = 0.$$

This exactly means that $f \perp u_0$, $u_0 = 1$ in $L_2(G)$. Hence, we have the self-adjoint operator A. ■

It is worth mentioning that the scheme we described in our preliminaries is applicable to general symmetric differential operator $A \leq 0$ associated with the corresponding functional spaces $W(G) \supseteq D_A$ and $W(G \times I)$ such that

$$\|u\|^2_{W(G\times I)} \asymp \int_I \left[\left\| \frac{\partial}{\partial t} u_t \right\|^2_{L_2(G)} + \|u_t\|^2_{W(G)} \right] dt, \qquad u \in W(G \times I); \tag{4.9}$$

recall, we consider a case of the elliptic operator A when (4.9) holds with $W(G) = W_2^p(G)$, $W(G \times I) = W_2^{(p,1)}(G \times I)$.

4.2. Semigroup approach to deterministic equations[4]

Let us consider equation (4.1)′ with the deterministic source $f \in L_2(G \times I)$, looking for the deterministic $u \in W(G \times I)$ of the form $u = u_t$, $t \in I$, with

$$\frac{\partial}{\partial t} u_t = A u_t + f_t, \tag{4.10}$$

where the corresponding $f_t \in L_2(G)$ is well-defined on $I = (t_0, t_1)$ only almost everywhere (a.e.) representing $f = f_t$, $t \in I$. The question is on existence of a solution with $u_t \in D_A$ from the domain of the self-adjoint operator $A: D_A \to L_2(G)$. We would like to show that

$$u_t = \int_{t_0}^{t} e^{A(t-s)} f_s \, ds, \qquad t \in I, \tag{4.11}$$

for any $f \in L_2(G \times I)$ gives the solution $u = u_t$, $t \in I$, of this equation with

$$u_{t_0} = 0, \qquad u_t \in D_A \text{ a.e.} \tag{4.12}$$

such that

$$\int_I \left[\left\| \frac{\partial}{\partial t} u_t \right\|^2_{L_2(G)} + \|u_t\|^2_{W(G)} \right] dt \leq C \|f\|^2_{L_2(G \times I)}. \tag{4.13}$$

According to (4.9), this formula implies

$$u \in W(G \times I).$$

To show this we apply a complete orthonormal system of eigenfunctions $\varphi_\lambda \in D_A$ of the operator A:

$$A\varphi_\lambda = \lambda \varphi_\lambda$$

with all $\lambda \leq 0$. This simplifies the representation (4.11) which takes the form

$$f_t = \sum_\lambda \delta_\lambda(t) \varphi_\lambda, \qquad t \in I,$$

with

$$\sum_\lambda \int_I |\delta_\lambda(t)|^2 \, dt = \int_I \|f_t\|^2_{L_2(G)} \, dt = \|f\|^2_{L_2(G \times I)}$$

[4] For the related general material see, for example, H. Tanabe: Equations of Evolutions, Pitman, London, 1979.

as

$$u_t = \sum_\lambda \left[\int_{t_0}^{t} e^{\lambda(t-s)} \delta_\lambda(s)\, ds \right] \varphi_\lambda, \qquad t \in I.$$

And one can easily verify (4.10)–(4.13) with the help of an elementary inequality

$$\int_{t_0}^{t_1} [|u'(t)|^2 + |\lambda u(t)|^2]\, dt \le \int_{t_0}^{t_1} |u(t)|^2\, dt$$

for

$$u(t) = \int_{t_0}^{t} e^{\lambda(t-s)} \delta(s)\, ds$$

with $\lambda \le 0$ which results from the differential equation

$$u'(t) - \lambda u(t) = \delta(t)$$

according to

$$\int_{t_0}^{t_1} |\delta|^2\, dt = \int_{t_0}^{t_1} |u' - \lambda u|^2\, dt$$

$$= \int_{t_0}^{t_1} |u'|^2\, dt - \lambda \int_{t_0}^{t_1} 2uu'\, dt + \int_{t_0}^{t_1} |\lambda u|^2\, dt$$

$$= \int_{t_0}^{t_1} [|u'|^2 + |\lambda u|^2]\, dt - \lambda u(t_1)^2.$$

Namely, for all

$$u_\lambda(t) = \int_{t_0}^{t_1} e^{\lambda(t-s)} \delta_\lambda(s)\, ds$$

with

$$u'_\lambda(t) = \lambda u_\lambda(t) + \delta_\lambda(t)$$

we have

$$\int_{t_0}^{t_1} \sum_\lambda [|u'_\lambda(t)|^2 + |\lambda u_\lambda(t)|^2]\, dt$$

$$\leq \int_{t_0}^{t_1} \sum_{\lambda} |\delta_\lambda(t)|^2 \, dt = \int_{t_0}^{t_1} \|f_t\|^2_{L_2(G)} = \|f\|^2_{L_2(G\times I)},$$

which implies

$$\sum_{\lambda} [|u'_\lambda(t)|^2 + |\lambda u_\lambda(t)|^2] < \infty \text{ a.e.}$$

since with well-defined

$$\sum_{\lambda} u_\lambda(t)\varphi_\lambda = Au_t, \qquad \sum_{\lambda} u'_\lambda(t)\varphi_\lambda = Au_t + f_t = u'_t$$

in $L_2(G)$ we obtain $u_t \in D_A$ a.e.,

$$u_t = \int_{t_0}^{t} u'_s \, ds$$

as a vector function of t in the $L_2(G)$-space, and (4.13) results from

$$\int_{t_0}^{t_1} \Big[\Big\|\frac{\partial}{\partial t} u_t\Big\|^2_{L_2(G)} + \|Au_t\|^2_{L_2(G)}\Big] \, dt \leq \|f\|^2_{L_2(G\times I)}$$

with $\partial u_t/\partial t = u'_t$, $t \in I$, according to (4.4), (4.9).

The formula (4.11) gives us the solution $u \in \mathbf{W}(G \times I)$ of the deterministic equation (4.10) with the zero boundary conditions of the type $X^+(\Gamma)$ indicated in (4.6), (4.6)′, which are equivalent to (4.12), namely according to (4.5) for $u_t \in D_A$ a.e. we have

$$(x \times \delta, u) = \int_I \delta(t)(x, u_t) \, dt = 0, \qquad x \in X^+(\partial G)$$

with $\delta \in C_0^\infty(I)$, and this equality for $u_t \in W(G)$ a.e. implies $(x, u_t) = 0$ with a countable complete system of $x \in X_A(\partial G)$.

Let us show that equation (4.10) has a *unique* solution $u \in W(G\times I)$ with the conditions (4.12).

The generalized equation (4.10) implies

$$\int_I \delta'(t) \Big[(\varphi, u_t) - \int_{t_0}^{t} (A\varphi, u_s) \, ds - \int_{t_0}^{t} f_s \, ds\Big] \, dt = 0, \qquad \varphi \in C_0^\infty(G),$$

for all $\delta \in C_0^\infty(I)$; hence

$$(\varphi, u_t) - \int_{t_0}^{t} (A\varphi, u_s)\, ds - \int_{t_0}^{t} (\varphi, f_s)\, ds = C, \qquad C = (\varphi, u_{t_0})$$

and in the case of (4.12), we have

$$(\varphi, u_t) = \left(\varphi, \int_{t_0}^{t} Au_s\, ds + \int_{t_0}^{t} f_s\, ds\right), \qquad \varphi \in C_0^\infty(G),$$

that is

$$u_t = \int_{t_0}^{t} Au_s\, ds + \int_{t_0}^{t} f_s\, ds, \qquad t_0 \le t \le t_1, \tag{4.14}$$

as a vector function in the $L_2(G)$-space. Considering the homogeneous equation (4.10) with $f = 0$ we obtain from (4.14) a function $\|u_t\|_{L_2}^2$, $t_0 \le t \le t_1$, which is absolutely continuous with a derivative

$$\frac{d}{dt}\|u_t\|_{L_2}^2 = 2(u_t, u_t') = 2(u_t, Au_t) \le 0 \text{ a.e.}$$

and $u_t \equiv 0$ in the case of (4.12) with $u_{t_0} = 0$.

Thus, the formula (4.11) describes the unique solution $u \in W(G \times I)$ of the equation (4.10) in the cylinder $S = G \times I$ with the *zero* boundary conditions of the type $X^+(\Gamma)$ indicated in (4.6), (4.6)′ as

$$(x \times \delta, u) = 0, \qquad x \in X^+(\partial G), \qquad \delta \in C_0^\infty(I)$$

on the cylinder-side boundary $\partial G \times I$ and

$$(\varphi \times \delta_{t_0}, u) = 0, \qquad \varphi \in C_0^\infty(G)$$

on its base boundary $G \times \{t_0\}$. This statement is valid for any $f \in L_2(G \times I)$ in (4.10).

In the framework of our general scheme with the test functions space $X(G \times I)$ employed in equation (4.10) in its generalized form

$$(\mathcal{L}^*\varphi, u) = (\varphi, f), \qquad \varphi \in C_0^\infty(G \times I)$$

with $\mathcal{L}^* = -\partial/\partial t - A$, this result implies the equivalence

$$\|\mathcal{L}^*\varphi\|_{X(G\times I)} \asymp \|\varphi\|_{L_2(G\times I)} \tag{4.15}$$

and the direct sum representation

$$X(G \times I) = \mathcal{L}^* L_2(G \times I) + X^+(\Gamma) \tag{4.16}$$

with the boundary *subspace* $X^+(\Gamma)$ generated by the test functions (4.6), (4.6)′ on the boundary $\Gamma = \partial S$ of the cylinder $S = G \times I$.

4.3. Stochastic equations[5]

According to (4.15), (4.16), we have the following result.

THEOREM. *The stochastic evolution equation* (4.1), (4.1)′ *with arbitrary given stochastic boundary conditions* (4.3) *of the type* $X^+(\Gamma)$ *indicated in* (4.6), (4.6)′ *has a unique solution* $u = \xi \in \mathbf{W}(G \times I)$.

Our initial interest concerns the semigroup formula (4.2) with stochastic source η_{ds} of the white noise type which, contrary to the deterministic case, gives

$$\xi_t = \int_{t_0}^{t} e^{A(t-s)}\, \eta_{ds}$$

far away from the domain D_A of the self-adjoint operator A.

One can observe, for example, that for the Gaussian white noise η_{ds}, *formula* (4.2) *with the general self-adjoint operator* $A \leq 0$ *gives* $\xi_t \in D_A$ *if only* A *is of the Hilbert-Schmidt type.* This just follows from the fact that the generalized Gaussian field

$$A\xi_t = (x, A\xi_t) \equiv (Ax, \xi_t), \qquad x \in C_0^\infty(G),$$

is represented by $A\xi_t \in L_2(G)$ if and only if its correlation operator R_A

$$(x, R_A x) = \mathrm{E}(Ax, \xi_t)^2, \qquad x \in C_0^\infty(G),$$

is *nuclear.* Namely, we have

$$R_A = AR_tA,$$

where

$$R_t = \frac{1}{2} \int_0^{2(t-t_0)} e^{As}\, ds$$

is the correlation operator of ξ_t

$$R_A = \frac{1}{2} A \int_0^{2(t-t_0)} e^{As} A\, ds = \frac{1}{2} A \int_0^{2(t-t_0)} \frac{d}{ds} e^{As}\, ds = \frac{1}{2} A[e^{2A(t-t_0)} - 1]$$

---

[5]For the related general material see, for example, B. L. Rozovskii, Stochastic Evolution Equations, Kluwer Academic Publishers, Dordrecht-Boston-London, 1990. Boundary problems were employed in S. Albeverio and Yu.A. Rozanov, On semigroup approach to stochastic evolution equations, Theory Prob. and Appl., 1993, vol. 38, N 1, 1–23.

and from spectral representation of the self-adjoint operator $A \le 0$, we have

$$R_A = \int_{-\infty}^{0} \frac{1}{2}\lambda[e^{\lambda(t-t_0)} - 1]\, dE_\lambda$$

which shows that R_A is nuclear if and only if the spectral measure dE_λ is discrete and the corresponding eigenvalues λ are such that

$$\sum_\lambda \lambda[e^{+\lambda(t-t_0)} - 1] \asymp \sum \lambda^2 < \infty$$

which describes A as the Hilbert–Schmidt operator.

Nevertheless, for our stochastic case the following result holds true.

THEOREM. *The semigroup formula* (4.2) *gives the unique solution* $u = \xi \in W(G \times I)$ *of the equations* (4.1), (4.1)′ *in the cylinder* $S = G \times I$ *with the zero boundary conditions* (4.3) *of the type* $X^+(\Gamma)$ *indicated in* (4.6), (4.6)′.

Let us prove it. We know that the unique solution $u = \xi \in W(G \times I)$ with zero boundary conditions does exist. And, according to the general solution formula associated with (4.3), all its values are in the linear closure of random variables

$$h = (g, h) = \int_I g_s\, \eta_{ds}, \qquad g \in L_2(G \times I).$$

Taking for $\xi = \xi_t$, $t \in I$, as

$$(x, \xi_t) = \int_{t_0}^{t} (Ax, \xi_t)\, ds + \int_{t_0}^{t} x\, \eta_{ds}, \qquad x \in C_0^\infty(G),$$

the mean value

$$(x, u_t) = \mathrm{E}(x, \xi_t)h$$

we obtain

$$(x, u_t) = \int_{t_0}^{t} (Ax, u_s)\, ds + \int_{t_0}^{t} (x, f_s)\, ds$$

with $f = Rg$, where R is a bounded correlation operator in $L_2(G \times I)$ of the generalized random field $(g, h) = \int_I g_s\, \eta_{ds}$, $g \in L_2(G \times I)$, generated by the stochastic source. Hence this $u = u_t$, $t \in I$, is the unique solution $u \in W(G \times I)$ of the corresponding deterministic equation (4.10), and according to the zero boundary conditions is nothing else but

$$u_t = \int_{t_0}^{t} e^{A(t-s)} f_s \, ds, \qquad t \in I.$$

Taking now a similar mean value for $\xi = \xi_t$, $t \in I$, represented by the semigroup formula (4.2), we obtain the same

$$u_t = (x, u_t) = \mathrm{E}(x, \xi_t)_h = \mathrm{E}\Big[\int_{t_0}^{t} e^{A(t-s)} x \, \eta_{ds}\Big]\Big[\int_{t_0}^{t} g_s \, \eta_{ds}\Big] =$$

$$= \Big(x, \int_{t_0}^{t} e^{A(t-s)} f_s \, ds\Big), \qquad x \in C_0^\infty(G).$$

Hence, the former solution $u = \xi \in \mathbf{W}(G \times I)$ is exactly

$$\xi_t = \int_{t_0}^{t} e^{A(t-s)} \, \eta_{ds}, \qquad t \in I.$$

The proof is over.

Note, the general solution of equations (4.1), (4.1)′ is that $\xi \in \mathbf{W}(G \times I)$ described above plus the solution $u \in \mathbf{W}(G \times I)$ of the homogeneous equation

$$\frac{\partial u}{\partial t} = Au$$

with the general stochastic boundary conditions (4.3) given on the boundary subspace $X^+(\Gamma)$, generated by the test functions (4.6), (4.6)′.

II.5. Homogeneous Equations

5.1. General solution and its approximations

Let us recall that in our general scheme (2.1)–(2.10) for

$$\mathcal{L}u = f$$

regarded as the generalized differential equation

$$(\mathcal{L}^* g, u) = (g, f), \qquad g \in F(S) = [C_0^\infty(S)]$$

in the region $S \subseteq T$ with the nondegenerate operator

$$\mathcal{L}^*\colon F(S) \to X^-(S) = \mathcal{L}^* F(S)$$

we can set arbitrary boundary conditions

$$(x, u) = (x, u^+), \qquad x \in X^+(\Gamma)$$

on any boundary subspace $X^+(\Gamma)$ in the direct sum

$$X(S) = [C_0^\infty(S)] = \mathcal{L}^* F(S) + X^+(\Gamma),$$

and the unique solution $u \in \mathbf{W}(S)$ can be described by the formula (2.16) as

$$u = (x, u) = (g, f) + (x^+, u^+), \qquad x \in X(S),$$

according to the corresponding direct sum representation $x = \mathcal{L}^* g + x^+$ with $g \in F(S)$, $x^+ \in X^+(\Gamma)$. Obviously, here

$$u = (x, u) = (g, f), \qquad x \in X(S),$$

gives a solution with the *zero* boundary conditions and

$$u = (x, u) = (x^+, u^+), \qquad x \in X(S),$$

represents the solution of the *homogeneous equation*

$$\mathcal{L} u = 0 \tag{5.1}$$

in the region S.

Taking a *complete system* $\{x_k\}$ in the boundary test functions subspace $X^+(\Gamma)$, we can set boundary conditions as

$$(x_k, u) = (x_k, u^+), \qquad x_k \in X^+(\Gamma). \tag{5.2}$$

Let this $\{x_k\}$ be a *basis* such that any $x \in X^+(\Gamma)$ can be represented as a strong limit

$$x = \lim_{n\to\infty} \sum_{k \le n} c_k x_k = \sum_k c_k x_k$$

with

$$\sum_k |c_k|^2 \asymp \|x\|_X^2, \qquad x \in X^+(\Gamma),$$

where the coefficients

$$c_k = (x_k, u_k)$$

are to be determined by means of the *dual system* $\{u_k\}$ of functions $u_k \in W(S)$ in the dual Hilbert space $W(S) = X(S)^*$,

$$(x_j, u_k) = \begin{cases} 1, & j = k, \\ 0, & j \neq k. \end{cases} \tag{5.3}$$

The dual system $\{u_k\}$ can be taken such that every $u_k \in W(S)$ represents a linear continuous functional on $X(S)$ which gives $(x_k, u_k) = 1$ on the corresponding element $x_k \in X^+(\Gamma)$ and zero on the other x_j, $j \neq k$, plus on the direct complement $\mathcal{L}^* F(S) = X(S) - X^+(\Gamma)$. In this way, we obtain the dual system $\{u_k\}$, where every $u = u_k \in W(S)$

$$(\mathcal{L}^* g, u_k) = 0, \qquad g \in F(S) = [C_0^\infty(S)],$$

is a deterministic solution of the deterministic boundary problem (5.1), (5.2) with the boundary conditions given in the form (5.3) by means of the basis $\{x_k\}$ in the boundary test functions subspace $X^+(\Gamma)$. The deterministic solutions $u_k \in W(S)$,

$$\mathcal{L} u_k = 0$$

in the region S, are expected to be "nice" functions depending on properties of the differential operator $\mathcal{L}$.

Any stochastic boundary conditions (5.2) can be set in the form

$$(x_k, u) = \xi_k, \qquad x_k \in X^+(\Gamma), \tag{5.4}$$

with random variables $\{\xi_k\}$ such that

$$\mathrm{E}\Big|\sum_k c_k \xi_k\Big|^2 \leq C \sum_k |c_k|^2$$

for all its linear combinations. The corresponding stochastic sample $u^+ \in \mathbf{W}(S)$ in (5.2) can be described as

$$(x, u^+) = \sum_k c_k \xi_k$$

for all linear combinations $x = \sum_k c_k x_k$. Indeed, this linear function $u^+ = (x, u^+)$ of x is meansquare continuous with respect to the norm $\|x\|_X$, because

$$\mathrm{E}|(x, u^+)|^2 = \mathrm{E}\left|\sum_k c_k \xi_k\right|^2 \le C \sum_k |c_k|^2 \asymp C\|x\|_X^2.$$

Of course, for any $u \in \mathbf{W}(S)$,

$$(x, u) = \sum_k c_k \xi_k$$

with $\xi_k = (x_k, u)$ for all linear combinations $x = \sum_k c_k x_k$, we have

$$\mathrm{E}|(x, u)|^2 = \mathrm{E}\left|\sum_k c_k \xi_k\right|^2 \le C\|x\|_X^2 \asymp C \sum_k |c_k|^2.$$

Obviously, in the direct sum representation $x = \mathcal{L}^* g + x^+$ of $x \in X(S)$ with $g \in F(S) = [C_0^\infty(S)]$, $x^+ \in X^+(\Gamma)$, we have

$$x^+ = \sum_k c_k \xi_k, \qquad \sum_k |c_k|^2 < \infty,$$

with the coefficients

$$c_k = (x^+, u_k) = (x, u_k).$$

Let us recall that $(\mathcal{L}^* g, u_k) \equiv 0$. Hence, for the solution $u \in \mathbf{W}(S)$ of the boundary problem (5.1), (5.4) we have

$$(x, u) = \left(x, \sum_k \xi_k u_k\right)$$

as a mean square limit

$$(x, u) = (x^+, u) = \lim_{n \to \infty} \left(\sum_{k \le n} c_k x_k, u\right) =$$

$$= \lim_{n \to \infty} \sum_{k \le n} (x, u_k) \xi_k = \lim_{n \to \infty} \left(x, \sum_{k \le n} \xi_k u_k\right).$$

We see that this solution is the limit

$$u = \lim_{n \to \infty} u^{(n)} \tag{5.5}$$

of the functions

$$u^{(n)} = \sum_{k \le n} \xi_k u_k$$

which are linear combinations of the deterministic function $\{u_k\}$ with the random coefficients $\{\xi_k\}$ given in the boundary conditions (5.4), and according to

$$\left\| x^+ - \sum_{k \le n} c_k x_k \right\|_X^2 \asymp \sum_{k>n} |c_k|^2 \to 0, \qquad n \to \infty,$$

the corresponding approximation $u^{(n)}$ to this solution is such that

$$\mathrm{E}|(x,u) - (x,u^{(n)})|^2 \le C \sum_{k>n} |c_k|^2, \qquad x \in X(S) \tag{5.6}$$

for $x^+ = \sum_k c_k x_k \in X^+(\Gamma)$.

Thus, we obtain the following result.

THEOREM. *The general boundary problem* (5.1), (5.4) *has a unique solution* $u \in \mathbf{W}(S)$ *which is represented by the aproximate formula* (5.5) *with the approximation rate indicated in* (5.6).

5.2. Hyperellipticity, extension of solutions, and negligible singularities

For the so-called hyperelliptic operator $\mathcal{L}$ any generalized solution of equation (5.1) is an infinitely differential function $u = u(t)$, $t \in S$, in the region S. Dealing with the general equation (5.1), let us call its deterministic solutions the *$\mathcal{L}$-functions.*

We are mainly interested in $u = \xi \in \mathbf{W}(S)$ which arises as solutions of the stochatic boundary problems (5.1), (5.2), and these generalized random functions turn out to have much worse "analytical" properties than the deterministic $u \in W(S)$.

Nevertheless, any stochastic solution $u = \xi \in \mathbf{W}(S)$ of the homogeneous equation (5.1) can be regarded as an $\mathcal{L}$-function in the sense that *all of its realization ξ_ω being Schwartz distributions*

$$\xi_\omega = (\varphi, \xi_\omega), \qquad \varphi \in C_0^\infty(S)$$

satisfy equation (5.1),

$$\mathcal{L}\xi_\omega = 0$$

in the region S, that is all realizations are the $\mathcal{L}$-functions (to be more precise there always exists an *equivalent modification* of the generalized

$$\xi = (\varphi, \xi), \qquad \varphi \in C_0^\infty(S),$$

with all its realizations ξ_ω being $\mathcal{L}$-harmonic functions in S).

It seems worth mentioning that the random $\mathcal{L}$-functions being smooth nice (analytic, say) in the region S could have extremely *chaotic* behaviour approaching the boundary $\Gamma = \partial S$. For example, in the scheme with the Laplace operator $\mathcal{L} = \Delta$ in the corresponding Sobolev space $\mathbf{F} = W = \overset{\circ}{W}{}_2^1$, we get *harmonic* functions $u \in \mathbf{W}(S) = \mathbf{W}_2^1(S)$ in the region S with generalized boundary values (x, u), $x \in X(\Gamma)$, on the boundary $\Gamma = \partial S$ which are nearly as much chaotic as the *white noise* with realizations on the boundary Γ being no better than from $W_2^{-q}(\Gamma)$, $q > \frac{d-1}{2}$. ■

In our general scheme we deal with $u \in \mathbf{W}(S)$ regarded as the restriction in the region S of some $u \in \mathbf{W}$ in an appropriate region $T \supseteq S$, and therefore, any generalized random function $u \in \mathbf{W}(S)$ can be extended to $u \in \mathbf{W}(S_0)$ in any region S_0

$$S \subseteq S_0 \subseteq T.$$

Suppose that we deal with the boundary problem (5.1), (5.2) in the region S of rather complicated structure; then it seems natural to try to solve an appropriate boundary problem in some better $S_0 \supseteq S$, in order to obtain the solution $u \in \mathbf{W}(S)$ we are interested in as the restriction of the corresponding solution $u \in \mathbf{W}(S_0)$ in the region S_0.

Let us take the following S,

$$S = S_0 \setminus \gamma,$$

where a closed set γ is some kind of *singular* set. Let us call γ the *negligible singular set* in the region S_0 if any $\mathcal{L}$-function $u \in \mathbf{W}(S)$ in the region $S = S_0 \setminus \gamma$ can be uniquely extended to an $\mathcal{L}$-function $u \in \mathbf{W}(S_0)$ in S_0. In the case of the negligible γ we do not need to worry any more, since any boundary problem (5.1), (5.2) in $S = S_0 \setminus \gamma$ is just the problem in S_0. The question arises, as to how we can characterize such a case in the framework of our general scheme with the differential operator $\mathcal{L}$ in the corresponding functional space $F = [C_0^\infty(T)]$.

Taking into account the fact that equation (5.1) in S_0 is served by the test functions subspace $[\mathcal{L}^* C_0^\infty(S_0)]$ and that the $\mathcal{L}$-function $u \in \mathbf{W}(S_0)$ is defined as

$$(\mathcal{L}^*\varphi, u) \equiv 0, \qquad \varphi \in F(S_0)[C_0^\infty(S_0)]$$

we see that the singular set γ is negligible if and only if a closure of $\mathcal{L}^* C_0^\infty(S_0)$ in the test functions space coincides with the closure of $\mathcal{L}^* C_0^\infty(S_0 \setminus \gamma)$

$$[\mathcal{L}^* C_0^\infty(S_0 \setminus \gamma)] = [\mathcal{L}^* C_0^\infty(S_0)].$$

Thus, for the nondegenerate operator $\mathcal{L}^*$ on $F(S_0) = [C_0^\infty)] \subseteq F$ we obtain the following result: *the singular set γ is negligible in S_0 if and only if*

$$[C_0^\infty(S_0 \setminus \gamma)] = [C_0^\infty(S_0)]. \tag{5.7}$$

in the corresponding functional space $\mathbf{F}$.

In particular, we see that in our scheme with the differential operator $\mathcal{L}$ in $F = L_2(T)$ any set γ of the *zero* Lebesgue measure is negligible.

For illustration, let us consider $\mathcal{L} = d/dt$ in $F = L_2(T)$ on $T \subseteq R^1$ and take the interval $S_0 = (t_0, t_1)$, $[S_0] \subseteq T$, with the set $\gamma \in S_0$ consisting of a single point; here for $S = S_0 \setminus \gamma$ we have our $W(S)$ regarded as the restriction of the Sobolev space $W(S_0) = W_2^1(S_0)$, and therefore, any $\mathcal{L}$-function $u = u(t)$, $t \in S$, being constant on the intervals $t_0 < t < \gamma$, $\gamma < t < t_1$, and *continuous* on $S_0 = (t_0, t_1)$, is just the *constant* on S_0. In the case of $\mathcal{L} = \mathcal{L}^* = \mathcal{P} = -d^2/dt^2$ in the corresponding Sobolev space $F = W = \overset{\circ}{W}{}_2^1(T)$, the singular point $\gamma \in S_0$ on the interval $S_0 = (t_0, t_1)$, $[S_0] \subseteq T$, is not negligible, because a broken line function $u(t)$, $t \in S = S_0 \setminus \gamma$, with different linear parts on the intervals $t_0 < t < \gamma$, $\gamma < t < t_1$, is $\mathcal{L}$-function in S but its unique extension $u \in W_2^1(S_0)$ on the interval S_0 is not.

In trying to solve the boundary problem in the region $S = S_0 \setminus \gamma$ by means of the corresponding boundary problem in S_0, we encounter a question concerning the appropriate $f = \mathcal{L}u$ in the region S_0 ($f = 0$ in S) and the appropriate boundary conditions on $\Gamma_0 = \partial S_0$ for $u \in W(S_0)$, which is supposed to be an extension in S_0 of the solution $u \in \mathbf{W}(S)$ we are looking for.

For example, let us consider our scheme with the differential operator $\mathcal{L}$ in $F = L_2(T)$ when $\mathcal{L}^*$ is not degenerate on $L_2(T)$, having in mind the boundary problem (5.1), (5.2) in S with boundary conditions on $\Gamma = \partial S$ of the type

$$X^+(\Gamma) = X_0^+(\Gamma)$$

which are of special interest to us (cf. the known forecast problem in Section 3.2). Recall, these boundary conditions with the stochastic sample $u^+ \in \mathbf{W}$ can be set as

$$(x, u) = (g^+, f), \qquad x = \mathcal{L}^* g^+ \in X_0^+(\Gamma)$$

by means of $f = \mathcal{L}u^+$ *outside* S, because we deal here with the boundary test functions $x = \mathcal{L}^* g^+$ which correspond to $g^+ \in L_2(T)$, $g^+ = 0$ in S and $\mathcal{L}^* g^+ = 0$ in the complementary region $S_+ = T \setminus [S]$ (see (2.6), (2.9)).

Any $f = \mathcal{L}u^+$, given outside S as a generalized random function

$$f = (g, f), \qquad g \in L_2(S_+),$$

can be extended on all $g \in L_2(T)$ in a way that

$$(g, f) = 0, \qquad g \in L_2(S)$$

in the region S. By means of this extension we can set the corresponding boundary problem in S_0 as

$$\mathcal{L}u = f \tag{5.8}$$

in the region S_0,

$$(x, u) = (g^+, f), \qquad x = \mathcal{L}^* g^+ \in X_0^+(\Gamma_0), \tag{5.9}$$

on its boundary $\Gamma_0 = \partial S_0$, where all $x = \mathcal{L}^* g^+$ corresponds to $g^+ \in L_2(T)$, $g^+ = 0$ in S_0 and $\mathcal{L}^* g^+ = 0$ in the complementary region. Obviously, a unique solution $u \in \mathbf{W}(S_0)$ of the boundary problem (5.8), (5.9) can be extended to $u \in \mathbf{W}$ as

$$(x, u) = (g, f), \qquad x = \mathcal{L}^* g, \qquad g \in L_2(T),$$

so we have $\mathcal{L}u = 0$ in the region S and $(x, u) = (g, f)$ for all $x = \mathcal{L}^* g$, $g = 0$ in S and $\mathcal{L}^* g = 0$ in S_+, with the *initially given* $f = \mathcal{L}u^+$ outside S. Thus, the restriction of this solution $u \in \mathbf{W}(S_0)$ in the region $S \subseteq S_0$ represents a solution of the initial boundary problem in S.

Moreover, when the singular set γ is "small" we can take

$$\mathcal{L}u_0 = 0$$

in S_0 as an approximation of the equation (5.8) with "small"

$$f = wf, \qquad w = 1_\gamma = \begin{cases} 1, & t \in \gamma, \\ 0, & t \notin \gamma, \end{cases}$$

in the region S_0, and the solution $u_0 \in \mathbf{W}(S_0)$ of the corresponding boundary problem will be an approximation of the solution $u \in \mathbf{W}(S_0)$ of the boundary problem (5.8), (5.9) according to

$$\mathrm{E}|(x,u) - (x,u_0)|^2 \le C \int_\gamma |g(t)|^2 \, dt$$

with $g \in L_2(S_0)$,

$$\|g\|_{L_2}^2 = \int_{S_0} |g(t)|^2 \, dt < \|x\|_X^2,$$

in the direct sum representation $x = \mathcal{L}^* g + x^+$, $x^+ \in X_0^+(\Gamma)$, for all test functions $x \in X(S_0)$.

5.3. Limit of solutions in regions with increasingly complicated structure

Let us imagine that we are dealing with a random field in the region $T \subseteq R^d$ which arises as a result of some stochastic source in T *outside* the region $S \subseteq T$ and, in the framework of our scheme with some differential operator $\mathcal{L}$ in $F = L_2(T)$, so this field can be represented by the corresponding $u \in \mathbf{W}$ with

$$\mathcal{L}u = f$$

where a generalized random function $f = (\varphi, f)$, $\varphi \in C_0^\infty(T)$, is meansquare continuous with respect to $\|\varphi\|_{L_2}$, $f = 0$ in $S \subseteq T$. Thus, this field is represented in the region S by the unique solution $u \in \mathbf{W}(S)$ of the boundary problem (5.1), (5.2) with boundary conditions on $\Gamma = \partial S$ of the type we know.

In the case when it is difficult to solve this boundary problem arising in the region S of a complicated structure, such as

$$S = S_0 \setminus \gamma$$

with some kind of a singular set γ, it was suggested to consider the corresponding boundary problem in the region S_0 — see (5.8), (5.9). It seems natural to analyse the question on some limit behaviour of the solution $u = u_n \in \mathbf{W}(S_0)$ of this boundary problem for $S = S_n$ with increasingly complicated structure

$$S_n = S_0 \setminus \gamma_n, \tag{5.10}$$

depending on the given $\gamma = \gamma_n$, $n \to \infty$ (for example, $\gamma = \gamma_n$ is formed by a large number of small "granules" randomly distributed in the region S_0). To be more precise, in the suggested scheme, the question is about a strong/weak limit

$$u = \lim u_n$$

of the solutions $u_n \in \mathbf{W}(S_0)$ of the corresponding boundary problems (5.8), (5.9) in the region S_0 when $n \to \infty$.

For the singular set γ_n with the indicator $w_n = 1_{\gamma_n}$, equation (5.8) takes the following form

$$\mathcal{L}u_n = w_n f \tag{5.11}$$

in the region S_0, and the solution $u_n \in \mathbf{W}(S_0)$ of the boundary problem (5.8), (5.9) is such that

$$(x, u_n) = (g, w_n f) + (g^+, f)$$

for all our test functions $x \in X(S_0)$ with the corresponding direct sum representation $x = \mathcal{L}^* g + x^+$, $x^+ = \mathcal{L}^* g^+$. We see that only those $(g, w_n f)$ that represent the solution

$$(x, u) = (g, w_n f)$$

with *zero* boundary conditions on $\Gamma_0 = \partial S_0$, depend on γ_n,

$$(g, w_n f) = (w_n g, f), \qquad g \in L_2(S_0).$$

Obviously, we do have a strong limit $\lim u_n = u$ in the form of a mean square limit

$$\lim(x, u_n) = (x, u) \tag{5.12}$$

for all random variables (x, u_n), $x \in X(S_0)$, when we have a strong limit

$$\lim w_n g, \qquad g \in L_2(S_0)$$

in the space $L_2(S_0)$, and generally we cannot say more about (5.12) because the generalized random function $f = (\varphi, f)$ is characterized just by its mean square continuity with respect to $\varphi \in L_2(S_0)$. Moreover, it is easy to see that the strong limit

$$\lim w_n g = wg \tag{5.12$'$}$$

does exist for all $g \in L_2(S_0)$ if and only if there is a *limit set* γ

$$\gamma = \lim \gamma_n \tag{5.13}$$

such that the Lebesgue measure

$$\text{mes}\,[(\gamma_n \circ \gamma) \cap B] \to 0$$

for

$$\gamma_n \circ \gamma = (\gamma_n \setminus \gamma) \cup (\gamma \setminus \gamma_n)$$

and any bounded $B \subseteq S_0$; of course, in (5.12)$'$ we should take the indicator

$$w = 1_\gamma = \begin{cases} 1, & t \in \gamma, \\ 0, & t \notin \gamma. \end{cases}$$

One can verify this by taking into account the fact that $|w_n| < 1$, and $\lim w_n g$ exists for *all* $g \in L_2(S_0)$ if it does for a complete system of indicator functions $g = 1_B$ in $L_2(S_0)$. The strong limit (5.12)$'$ leads to the mean square limit

$$\lim(w_n g, f) = (wg, f) = (g, wf)$$

in (5.12) and provide us with the corresponding limit function

$$u = \lim u_n \in \mathbf{W}(S_0),$$

$$(x, u) = (g, wf) + (g^+, f)$$

for all our test functions $x \in X(S_0)$, $x = \mathcal{L}^* g + x^+$, $x^+ = \mathcal{L}^* g^+$. Hence, $u \in \mathbf{W}(S_0)$ is the unique solution of the boundary problem of the type (5.8), (5.9) with the corresponding limit equation

$$\mathcal{L}u = wf \tag{5.14}$$

in the region S_0.

Let us now consider the question concerning a *weak limit* $u = \lim u_n$ regarded as a limit of mean values

$$\lim \mathrm{E}[(x, u - u_n)h] = 0$$

for all our test functions $x \in X(S_0)$ and random variables h, $\mathrm{E}\|h\|^2 < \infty$.

Considering this question in the general setting, in the particular case of the deterministic $f \in L_2(S_0)$, we see that there must exist a limit of $(w_n g, f)$ which in the case of indicator functions $f = g = 1_B$ gives

$$\lim \text{mes}\,(\gamma_n \cap B) = \int_B w(t)\,dt. \tag{5.15}$$

Indeed, the limit $\lim \int_B w_n(t)\,dt$ for $w_n = 1_{\gamma_n}$ depends on $B \subseteq S_0$ as a measure which is absolutely continuous with respect to the Lebesgue measure and therefore is represented as in (5.15) with some density w,

$$0 \le w(t) \le 1, \qquad t \in S_0.$$

On the other hand, since $\|w_n\| \le 1$, condition (5.15) implies a limit

$$\lim(w_n g, f) = (wg, f)$$

for all indicator functions f, g and, moreover, for all $f, g \in L_2(S_0)$; thus, (5.15) implies a weak limit

$$\lim w_n g = wg$$

of $w_n g$ in the space $L_2(S_0)$ and we have our mean values limit

$$\lim \mathrm{E}[(w_n g - wg, f)h] = 0$$

in the case of generalized random functions $f = (\varphi, f)$, $\varphi \in L_2(S)$, which are meansquare continuous with respect to $\|\varphi\|_{L_2}$. We see that condition (5.15) implies our weak limit $\lim u_n = u$ which provides us with the solution $u \in \mathbf{W}(S_0)$ of equation (5.14) with the limit density function w from (5.15).

It is worth mentioning that *depending on the corresponding* γ_n, $n \to \infty$, *any function* w, $0 \le w \le 1$, *can appear in* (5.15). Indeed, let us take *disjoint* sets B_n^j, $j = 0, 1, \dots, n$, with the property that for any bounded $B \subseteq S_0$, there exist corresponding unions $B_n = \cup B_n^j$ such that

$$\lim_{n\to\infty} \text{mes}\,(B_n \circ B) = 0.$$

Taking

$$C_n^0 = \{t\colon w = 0\}, \qquad C_n^k = \left\{t\colon \frac{k-1}{n} < w \le \frac{k}{n}\right\}, \quad k = 1, \dots, n,$$

for arbitrary given w, $0 \le w \le 1$, we can apply $B_n = \cup A_n^{jk}$ with some other *disjoint* sets $A_n^{jk} = B_n^j \cap C_n^k$. Let us now take closed $\gamma_n^{jk} \subseteq A_n^{jk}$

$$\operatorname{mes} \gamma_n^{jk} = \frac{k}{n} \operatorname{mes} A_n^{jk} + o(n^{-2})$$

and set

$$\gamma_n = \bigcup_{j,k=0}^{n} \gamma_n^{jk}.$$

Obviously, there exists a limit

$$\lim_{n\to\infty} \operatorname{mes}(\gamma_n \cap B) = \lim_{n\to\infty} \sum_{j,k:A_n^{jk}\subseteq B_n} \frac{k}{n} \operatorname{mes} A_n^{jk}$$

$$= \lim_{n\to\infty} \int_{B_n} w(t)\, dt = \int_{B} w(t)\, dt.$$

Let us formulate our results for the scheme with *all* stochastic sources f which can possibly appear in (5.8), (5.9) a follows.

THEOREM. *The strong/weak limit of solutions* $u_n \in \mathbf{W}(S_0)$ *does exist if and only if condition* (5.13)/(5.15) *holds; the limit* $u = \lim u_n$ *is the unique solution of the boundary problem of the type* (5.8), (5.9) *considered with the corresponding limit equation* (5.14).

As an example, let us imagine that we are dealing with $S_n = S_0 \setminus \gamma_n$ in R^d, where

$$\gamma_n = \bigcup_k \gamma_n^k$$

is formed by randomly distributed small "granules" γ_n^k each being a ball of volume $\sigma^2 r^d$ with a radius

$$r = \left(\frac{\lambda}{n}\right)^{\frac{1}{d}}, \qquad \lambda = \text{const}, \qquad n \to \infty,$$

centred at random points $\tau_k \in S_0$, $k = 1, \dots, n$, which are independently distributed over S_0 with the same probability density $p(t)$, $t \in S_0$. Then *the weak limit condition* (5.15) *holds true with the limit function*

$$w(t) = 1 - e^{-\lambda\sigma^2 p(t)}, \qquad t \in S_0, \tag{5.16}$$

giving the limit equation (5.14).

Let us show this, assuming that $n = m^2$, $m = 1, 2, \ldots$, and that the probability density $p(t)$, $t \in S_0$, is continuous. For random variables

$$\mu_n = \operatorname{mes}(\gamma_n \cap B) = \int_B 1_{\gamma_n}(t)\,dt,$$

$$1_{\gamma_n}(t) = 1 - \prod_{k=1}^{n}(1 - 1_{\gamma_n^k}(t)), \qquad t \in S_0,$$

with indicators $1_{\gamma_n^k}$, $k = 1, \ldots, n$, of the random balls, we have a mean value

$$\mathrm{E}\mu_n = \int_B \Big[1 - \prod_k (1 - \mathrm{E}1_{\gamma_n^k}(t))\Big]\,dt,$$

where

$$\prod_{k=1}^{n}(1 - \mathrm{E}1_{\gamma_n^k}(t)) \to e^{-\lambda\sigma^2 p(t)}$$

for

$$\mathrm{E}1_{\gamma_n^k}(t) \sim \int_{|s-t|\le r} p(s)\,ds \sim p(t)\sigma^2 r^d = \frac{\lambda\sigma^2 p(t)}{n},$$

and thus

$$\lim_{n\to\infty} \mathrm{E}\mu_n = \int_B (1 - e^{-\lambda\sigma^2 p(t)})\,dt.$$

Besides, we have

$$\mathrm{E}\mu_n^2 = \iint_{B\times B} \mathrm{E}\Big[1 - \prod_j (1 - 1_{\gamma_n^j}(s))\Big]\Big[1 - \prod_k (1 - 1_{\gamma_n^k}(t))\Big]\,ds\,dt,$$

where, by taking integrals over all $(s,t) \in B \times B$, $|s-t| > r$, we deal only with independent $1_{\gamma_n^j}(s)$ and $1_{\gamma_n^k}(t)$, $j \neq k$, because

$$1_{\gamma_n^j}(s) 1_{\gamma_n^k}(t) = 0, \qquad j = k$$

and it is easy to see that

$$\lim_{n\to\infty} \mathrm{E}\mu_n^2$$

$$= \lim_{n\to\infty} \iint_{\substack{B\times B\\ |s-t|>r}} \mathrm{E}\Big[1 - \prod_j (1 - 1_{\gamma_n^j}(s))\Big]\,\mathrm{E}\Big[1 - \prod_k (1 - 1_{\gamma_n^k}(t))\Big]\,ds\,dt$$

$$= \int\limits_{B\times B} [1 - e^{-\lambda\sigma^2 p(s)}][1 - e^{-\lambda\sigma^2 p(t)}]\, ds\, dt$$

$$= \Big[\int\limits_B (1 - e^{-\lambda\sigma^2 p(t)})\, dt\Big]^2.$$

Hence,

$$\operatorname{Var}\mu_n = \mathrm{E}\mu_n^2 - (\mathrm{E}\mu_n)^2 \to 0.$$

Obviously, for $r^d = \lambda/n$, $n = m^2$, we have

$$\operatorname{Var}\mu_n = \iint\limits_{\substack{B\times B\\ |s-t|\le r}} \mathrm{E}1_{\gamma_n(s)}1_{\gamma_n(t)}\, ds\, dt$$

$$- \iint\limits_{\substack{B\times B\\ |s-t|\le r}} \mathrm{E}1_{\gamma_n(s)}\mathrm{E}1_{\gamma_n(t)}\, ds\, dt = O\Big(\frac{1}{m^2}\Big),$$

and with probability 1

$$\lim \mu_n = \int\limits_B (1 - e^{-\lambda\sigma^2 p(t)})\, dt.$$

Chapter III

Random Fields Associated with Partial Differential Equations

III.1. Some Characteristics of Probability

1.1. Mean values and correlations

In the framework of our general scheme considered in Chapter 2, we were discussing random fields ξ in the region $T \subseteq R^d$ which result from the stochastic source η according to the generalized differential equation

$$\mathcal{L}\xi = \eta \tag{1.1}$$

defined in T, or in the region $S \subseteq T$ equipped with boundary conditions on $\Gamma = \partial S$ of the form

$$(x, \xi) = (x, \xi^+), \qquad x \in X^+(\Gamma), \tag{1.2}$$

by means of the corresponding boundary test functions $x \in X^+(\Gamma)$ and appropriate boundary random field ξ^+.

The probability model (1.1), (1.2) seems to be most interesting when we can define the stochastic source η to be as much chaotic as the white noise and set the boundary conditions on the boundary $\Gamma = \partial S$ of the region S *independently* of this source in S. This requires in our scheme with the operator $\mathcal{L}$ in the L_2-space to have the nondegenerate operator

$$\mathcal{L}^*\colon\ L_2(S) = [C_0^\infty(S)] \to X(S)$$

and boundary conditions on the boundary subspace $X^+(\Gamma)$ of direct complement type

$$X^+(\Gamma) = X(S) - \mathcal{L}^* L_2(S)$$

in the test function space $X(S) = [C_0^\infty(S)]$ we encountered many times (see (2.18), Chapter 2).

One can imagine, for example, that we are actually interested in some deterministic boundary problem which is described by

$$\mathcal{L}u = f \tag{1.3}$$

in the region $S \subseteq T$ with the boundary conditions

$$(x, u) = (x, u^+), \qquad x \in X^+(\Gamma), \tag{1.4}$$

and some stochastic disturbance makes us to deal with the stochastic boundary problem (1.1), (1.2) represented by the corresponding η and ξ^+ with mean values

$$\mathrm{E}\eta = f, \qquad \mathrm{E}\xi^+ = u^+.$$

Then, of course, the solution $u \in W(S)$ we are interested in is the *mean value*

$$u = \mathrm{E}\xi$$

of the corresponding random field $\xi = \mathbf{W}(S)$. Therefore, starting with given $f = \mathrm{E}\eta$, $u^+ = \mathrm{E}\xi^+$, we can determine $u = \mathrm{E}\xi \in W(S)$ as a unique solution of the corresponding boundary problem (1.3), (1.4).

One might be interested in the *correlation operator* of the random field $\xi \in \mathbf{W}(S)$ which we consider in the case of $\mathrm{E}\xi = 0$, defined as

$$Q\colon\ X(S) \to W(S)$$

with

$$Qy = (\varphi, Qy) \equiv \mathrm{E}(\varphi, \xi)(y, \xi), \qquad \varphi \in C_0^\infty(S),$$

for all $y \in X(S)$; here $Qy \in W(S)$ will be called the *correlation function.*

It turns out, that with the given correlation of the random fields η in the regions S and ξ^+ on the boundary $\Gamma = \partial S$, we can get the correlation function

$$u = Qy$$

for any $y \in X(S)$ as a solution of the corresponding boundary problem (1.3), (1.4) associated with (1.1), (1.2). Recall, the stochastic source η in (1.1), represented by the generalized random field $\eta = (\varphi, \eta)$, $\varphi \in C_0^\infty(S)$, meansquare continuous with respect to $\|\varphi\|_F$ in the functional space F of our scheme, and it is well-defined on the corresponding $F(S) = [C_0^\infty(S)]$, $F(S) = L_2(S)$ or $F(S) = \overset{\circ}{W}(S)$. According to equation (1.1), for $y = \mathcal{L}^* g$, $g \in F(S)$ we have

$$(\mathcal{L}^*\varphi, Qy) = \mathrm{E}(\mathcal{L}^*\varphi, \xi)(\mathcal{L}^* g, \xi)$$
$$= \mathrm{E}(\varphi, \eta)(g, \eta) = (\varphi, Rg), \qquad \varphi \in C_0^\infty(S),$$

with the correlation operator R of the random field η, $\mathrm{E}\eta = 0$. We see that the correlation function $u = Qy$ satisfies the generalized differential equation (1.3) with $f = Rg$. In the case of zero boundary conditions (1.2) with $\xi^+ = 0$, *$u = Qy$ is a unique solution $u \in W(S)$ of equation* (1.3) *with*

$$f = Rg \tag{1.5}$$

and zero boundary conditions (1.4)

$$(x, u) = 0, \qquad x \in X^+(\Gamma).$$

Of course, for $y \in X^+(\Gamma)$ we get $Qy = 0$. In the case when the random field ξ results from equation (1.1) with $\eta = 0$ and the general boundary conditions (1.2), for $y = \mathcal{L}^* g$ we get $Qy = 0$, and for $y \in X^+(\Gamma)$ *the correlation function $u = Qy$ is a unique solution of equation* (1.3) *with $f = 0$ and boundary conditions* (1.4) *of the form*

$$(x, u) = (x, Q^+ y), \qquad x \in X^+(\Gamma), \tag{1.6}$$

with the correlation function $u^+ = Q^+ y$ of the boundary random field ξ^+,

$$(x, Q^+ y) = \mathrm{E}(x, \xi^+)(y, \xi^+) = \mathrm{E}(x, \xi)(y, \xi), \qquad x, y \in X^+(\Gamma).$$

These two examples show how to proceed with the correlation function of the random field ξ in general, when we have independent/noncorrelated η and ξ^+.

As an example, let us consider equation (1.1) with the zero boundary conditions (1.2) in our scheme with the differential operator $\mathcal{L} = \mathcal{L}^* = \mathcal{P} \geq 0$ and the white noise η on the corresponding space

$$F(S) = \overset{\circ}{W}(S) = [C_0^\infty(S)]$$

which is the closure of all $\varphi \in C_0^\infty(S)$ with respect to $\|\varphi\|_W = (\varphi, \mathcal{P}\varphi)^{\frac{1}{2}}$; as we know, the only boundary conditions (1.2) of the required type are $X^+(\Gamma) = X(\Gamma)$, and the zero boundary conditions for all deterministic $u \in W(S)$ mean that

$$u \in \overset{\circ}{W}(S) = [C_0^\infty(S)].$$

The correlation operator of the random field η is obviously $R = \mathcal{P}$, because

$$\mathrm{E}(\varphi, \eta)(g, \eta) = (\varphi, \mathcal{P}g), \qquad \varphi \in C_0^\infty(S),$$

with arbitrary $g \in F(S) = \overset{\circ}{W}(S)$, and the correlation function $u = Qy$ for $y = \mathcal{L}^* g$ is a unique solution $u \in \overset{\circ}{W}(S)$ of equation

$$\mathcal{L}u = Rg$$

with $\mathcal{L} = \mathcal{L}^* = \mathcal{P}$, $R = \mathcal{P}$ so we have

$$\mathcal{P}Q\mathcal{P}g = \mathcal{P}g, \qquad g \in \overset{\circ}{W}(S).$$

Hence, the correlation operator $Q\colon X(S) \to W(S)$ is

$$Q = \mathcal{P}^{-1} \tag{1.7}$$

on the subspace $\mathcal{P}\overset{\circ}{W}(S)$ and $Q = 0$ on its orthogonal complement $X(\Gamma) = X(S) - \mathcal{P}\overset{\circ}{W}(S)$. In particular, for $S = T$ we have $\mathcal{P}\overset{\circ}{W}(T) = X(T)$, thus the correlation operator is determined just by the formula (1.7).

The formula (1.7) with $P = \mathcal{L}^*\mathcal{L}$ holds true in our scheme with the general operator $\mathcal{L}$ in $F = L_2(T)$ and the white noise η on the $L_2(T)$-space as well; as we know, this case can be reduced to the scheme with the operator $P \geq 0$ in the corresponding $F = \overset{\circ}{W}(T)$.

1.2. Characteristic functional

The random field $\xi \in \mathbf{W}(S)$ described in (1.1) and (1.2) can be regarded as a solution of equation (1.1) with the *zero* boundary conditions (1.2) plus the solution of the *homogeneous* equation (1.1) with boundary conditions (1.2). Let us consider the first part which results from the stochastic source represented by some generalized random field η with independent values, $\mathrm{E}\eta = 0$.

One can imagine, for example, that there is some *Gaussian* source distributed in the region S as a Gaussian measure $\mu(dt)$, $t \in S$, with

$$\mathrm{E}\mu(dt) = 0, \qquad \mathrm{E}|\mu(dt)|^2 = F(dt)$$

plus Poisson type components $r\nu(dt\,dr)$ with $r \in R$, $R = R^1 \setminus \{0\}$, and the centered Poisson measure $\nu(dt\,dr)$ on the direct product $R \times S$

$$\mathrm{E}\nu(dt\,dr) = 0, \qquad \mathrm{E}|\nu(dt\,dr)|^2 = G(dt\,dr),$$

which together form the generalized random field $\eta = (\varphi, \eta)$, $\varphi \in C_0^\infty(S)$, with

$$(\varphi, \eta) = \int \varphi(t)\,\mu(dt) + \int\!\!\int \varphi(t) r\,\nu(dt\,dr) \tag{1.8}$$

where for random variables (φ, η) we have

$$\mathrm{E}(\varphi, \eta) = 0,$$

$$\mathrm{E}|(\varphi, \eta)|^2 = \int |\varphi(t)|^2\,F(dt) + \iint |\varphi|^2 r^2\,G(dr\,dt).$$

Dealing with real random variables (φ, η) one can take into account their *characteristic functional* $\mathrm{E}e^{i(\varphi,\eta)}$, $\varphi \in C_0^\infty(S)$, in the Lévy–Khinchine form

$$\log \mathrm{E}e^{i(\varphi,\eta)} = -\frac{1}{2}\int |\varphi(t)|^2\,F(dt) + \iint [e^{i\varphi(t)r} - 1 - i\varphi(t)r]\,G(dr\,dt). \tag{1.9}$$

Obviously, for the stochastic source described in (1.8), the random field $\xi \in \mathbf{W}(S)$ in (1.1) and (1.2) with the *zero* boundary conditions can be described as

$$(x, \xi) = \int g(t)\,\mu(dt) + \int\!\!\int g(t) r\,\nu(dr\,dt), \tag{1.9}$$

with

$$\log \mathrm{E}e^{i(x,\xi)} = -\frac{1}{2}\int |\varphi(t)|^2\,F(dt) + \int\!\!\int [e^{ig(t)r} - 1 - ig(t)r]\,G(dr\,dt)$$

according to the known direct sum representation

$$x = \mathcal{L}^* g + x^+, \qquad g \in F(S), \qquad x^+ \in X^+(\Gamma),$$

and $(x, \xi) = (\mathcal{L}^* g, \xi) = (g, \eta)$; recall that (g, η) is well-defined for all $g \in F(S) = [C_0^\infty(S)]$.

III.2. Forecast and Global Markov Property

2.1. Forecast problem

Let us consider our Probability model (1.1), (1.2) in some region $S = S_0$ representing a random field ξ by means of the generalized differential equation

$$\mathcal{L}\xi = \eta \tag{2.1}$$

in the region $S_0 \subseteq R^d$ with boundary conditions

$$(x, \xi) = (x, \xi^+), \qquad x \in X^+(\Gamma_0), \tag{2.2}$$

on the boundary $\Gamma_0 = \partial S_0$; recall that

$$\xi = (x, \xi), \qquad x \in X(S_0),$$

can be tested by

$$x \in X(S_0) = [C_0^\infty(S_0)]$$

as generalized functions (in an appropriate region $T \supseteq S_0$) with supports $\operatorname{supp} x \subseteq [S_0]$ in the closure of S_0.

The general *Forecast Problem* we are going to discuss concerns the conditional expectation

$$\hat{\xi} = \mathrm{E}(\xi/\mathcal{B})$$

being described as

$$\hat{\xi} = (x, \hat{\xi}) = \mathrm{E}[(x, \xi)/\mathcal{B}], \qquad x \in X(S_0),$$

with respect to some data $\mathcal{B}$. According to (2.1), (2.2), we obtain

$$(\mathcal{L}^*\varphi, \hat{\xi}) = \mathrm{E}[(\mathcal{L}^*\varphi, \xi)/\mathcal{B}] = \mathrm{E}[(\varphi, \eta)/\mathcal{B}] = (\varphi, \hat{\eta}), \qquad \varphi \in C_0^\infty(S_0),$$

with $\hat{\eta} = \mathrm{E}(\eta/\mathcal{B})$ and

$$(x, \hat{\xi}) = \mathrm{E}[(x, \xi)/\mathcal{B}] = \mathrm{E}[(x, \xi^+)/\mathcal{B}] = (x, \hat{\xi}^+), \qquad x \in X^+(\Gamma_0);$$

with $\hat{\xi}^+ = \mathrm{E}(\xi^+/\mathcal{B})$; thus, the forecast $\hat{\xi}$ *can be obtained as the unique solution* $\hat{\xi} \in \mathbf{W}(S_0)$ *of the equation*

$$\mathcal{L}\hat{\xi} = \hat{\eta} \tag{2.1$'$}$$

in the region S_0 *with the boundary conditions*

$$(x, \hat{\xi}) = (x, \hat{\xi}^+), \qquad x \in X^+(\Gamma_0), \tag{2.2$'$}$$

on $\Gamma_0 = \partial S_0$.

Obviously, the same result can be obtained with

$$\hat{\eta} = \widehat{\mathrm{E}}(\eta/\mathcal{B}), \qquad \hat{\xi}^{+} = \widehat{\mathrm{E}}(\xi^{+}/\mathcal{B})$$

for the so-called *linear forecast* $\hat{\xi} = \widehat{\mathrm{E}}(\xi/\mathcal{B})$ which is formed as the projection on the corresponding subspace of random variables generated the given $\mathcal{B}$; recall that in the Gaussian case, the linear forecast is the best one with the projection operator

$$\widehat{\mathrm{E}}(\cdot/\mathcal{B}) = \mathrm{E}(\cdot/\mathcal{B}). \quad \blacksquare$$

Let us consider our model (2.1), (2.2) with the stochastic source η, $\mathrm{E}\eta = 0$, in the region S_0 *independent* of the boundary conditions on $\Gamma_0 = \partial S_0$ given with the independent boundary sample $\xi^{+} \in \mathbf{W}(S_0)$.

For example, one can imagine that $\mathcal{B}$ in (2.1)$'$, (2.2)$'$ represents the boundary data given in the boundary conditions (2.2). Then the forecast $\hat{\xi}$ can be obtained as the unique solution $\hat{\xi} = u \in \mathbf{W}(S_0)$ of the homogeneous equation

$$\mathcal{L}u = 0 \tag{2.3}$$

in the region S_0 with the boundary conditions

$$(x, u) = (x, \xi), \qquad x \in X^{+}(\Gamma_0), \tag{2.4}$$

on the boundary $\Gamma_0 = \partial S_0$ given by the very data $\mathcal{B} = \{(x, \xi), x \in X^{+}(\Gamma)\}$.

Suppose that $\mathcal{B}$ represents some data concerning the random field ξ outside the region $S \subseteq S_0$ which include all boundary values

$$\mathcal{A}(\Gamma) = \{(x, \xi),\ x \in X(\Gamma)\} \tag{2.5}$$

on the boundary $\Gamma = \partial S$ and that we want to find out about ξ in S; for example, $\mathcal{B} = \mathcal{A}(S^c)$ represents all values

$$\mathcal{A}(S^c) = \{(x, \xi),\ \operatorname{supp} x \subseteq S^c\} \tag{2.6}$$

of the random field ξ with the test functions

$$x \in X(S_0), \qquad \operatorname{supp} x \subseteq S^c,$$

in the complement S^c of the region $S \subseteq S_0$.

We know that with given boundary values (2.5), the random field ξ in S can be identified as the unique solution $\xi \in \mathbf{W}(S)$ of the generalized differential equation

$$\mathcal{P}\xi = \eta \tag{2.7}$$

which is the same as (2.1) for $\mathcal{L} = \mathcal{L}^* = \mathcal{P} \geq 0$ in the corresponding space $F = W$ and with $P = \mathcal{L}^*\mathcal{L}$ and a *new* stochastic source η resulting from (2.1) after application of the operator $\mathcal{L}^*$ for the general $\mathcal{L}$ in $F = L_2$. Similarly to (2.1)′, (2.2)′, we can get $\hat{\xi}$ as the unique solution $\hat{\xi} \in \mathbf{W}(S)$ of the corresponding equation

$$\mathcal{P}\hat{\xi} = \hat{\eta}$$

in S with $\hat{\eta} = \mathrm{E}(\eta/\mathcal{B})$ and the Dirichlet type boundary conditions

$$(x, \hat{\xi}) = (x, \xi), \qquad x \in X(\Gamma),$$

given by means of the boundary data $\mathcal{A}(\Gamma) \subseteq \mathcal{B}$.

In our model (2.1), (2.2), let us assume that the stochastic source η is the Gaussian *white noise, independent* of the boundary conditions on the boundary $\Gamma_0 = \partial S_0$. Then the following result holds.

THEOREM. *The forecast $\hat{\xi}$ in the region $S \subseteq S_0$ with the given data $\mathcal{B} = \mathcal{A}(S^c)$ outside S can be obtained as the unique solution $\hat{\xi} = u \in \mathbf{W}(S)$ of the generalized Dirichlet problem*

$$\mathcal{P}u = 0 \tag{2.8}$$

in the region S,

$$(x, u) = (x, \xi), \qquad x \in X(\Gamma), \tag{2.9}$$

on the boundary $\Gamma = \partial S$.

Of course, this result is applicable to the conditional expectation

$$\hat{\xi} = \mathrm{E}(\xi/\mathcal{B})$$

in the case of any data $\mathcal{B}$ outside S containing the boundary values

$$\mathcal{A}(\Gamma) \subseteq \mathcal{B} \subseteq \mathcal{A}(S^c)$$

(see (2.5), (2.6)), for with

$$(x, u) = \mathrm{E}[(x, \xi)/\mathcal{A}(S^c)] = \mathrm{E}[(x, \xi)/\mathcal{A}(\Gamma)]$$

we have

$$(x, \hat{\xi}) = \mathrm{E}[(x, \xi)/\mathcal{B}] = \mathrm{E}\{\mathrm{E}[(x, \xi)/\mathcal{A}(S^c)]/\mathcal{B}\}$$
$$= \mathrm{E}\{\mathrm{E}[(x, \xi)/\mathcal{A}(\Gamma)]/\mathcal{B}\} = \mathrm{E}[(x, \xi)/\mathcal{A}(\Gamma)].$$

In the proof of (2.8), (2.9), we must take into account the fact that *the corresponding stochastic source* η *in equation* (2.7) *is the Gaussian white noise on the known* $\overset{\circ}{W}(S_0)$*-space with*

$$\mathrm{E}(\varphi, \eta) = 0, \qquad \mathrm{E}|(\varphi, \eta)|^2 = \|\varphi\|_W^2 = (\varphi, \mathcal{P}\varphi),$$
$$\varphi \in \overset{\circ}{W}(S) = [C_0^\infty(S_0)].$$

It is sufficient to prove that η in the region $S \subseteq S_0$ is *independent* of $\mathcal{B} = \mathcal{A}(S^c)$ *outside* S, because in such a case we would get $\hat{\eta} = 0$ in equation (2.7).

The given data $\mathcal{B} = \mathcal{A}(S^c)$ contain all boundary values present in the boundary conditions (2.2) on the boundary $\Gamma_0 = \partial S_0$ of the region $S_0 \supseteq S$, and by means of these boundary values we can form a component ξ^+ of our random field ξ as the unique solution $u = \xi^+ \in \mathbf{W}(S_0)$ of the homogeneous equation (2.1)

$$\mathcal{L}u = 0$$

in S_0, with

$$(x, u) = (x, \xi^+), \qquad x \in X^+(\Gamma_0),$$

on $\Gamma_0 = \partial S_0$. The component $\xi^- = \xi - \xi^+$, being the unique solution $\xi^- \in \mathbf{W}(S_0)$ of the equation

$$\mathcal{L}\xi^- = \eta$$

in S_0 with the *zero* boundary conditions

$$(x, \xi^-) = 0, \qquad x \in X^+(\Gamma_0),$$

on $\Gamma_0 = \partial S_0$, is *independent* of the component ξ^+, because η is independent of ξ^+ in the boundary conditions (2.2).

Let us first consider our scheme (2.1), (2.2) for $\mathcal{L} = \mathcal{L}^* = \mathcal{P} \geq 0$ in $F = W$ when we deal with the original stochastic source η in the region S_0 as the Gaussian white noise on the corresponding $F(S_0) = \overset{\circ}{W}(S_0) = [C_0^\infty(S_0)]$ and the boundary conditions given on the total boundary space $X^+(\Gamma_0) = X(\Gamma_0)$. We have $\xi^- \in \mathbf{W}(S_0)$ as

$$(x, \xi^-) = (g, \eta)$$

with the direct sum representation

$$x = \mathcal{L}^* g + x^+, \qquad g \in F(S_0) = \overset{\circ}{W}(S_0), \qquad x^+ \in X(\Gamma_0),$$

of all our test functions

$$x \in X(S_0) = [C_0^\infty(S_0)].$$

For $x = \mathcal{L}^* g + x$ with supports of x, x^+ and $\mathcal{L}^* g = \mathcal{P} g$ in the complement S^c we have

$$(\varphi, \mathcal{P} g) = \langle \varphi, g \rangle_W = 0, \qquad \varphi \in C_0^\infty(S);$$

hence

$$\mathrm{E}(\varphi, \eta)(x, \xi^-) = \mathrm{E}(\varphi, \eta)(g, \eta) = \langle \varphi, g \rangle_W = 0, \qquad \varphi \in C_0^\infty(S),$$

and therefore the Gaussian variables (φ, η), $\varphi \in C_0^\infty(S)$, are independent of (x, ξ^-), $\operatorname{supp} x \subseteq S^c$, and we see that the stochastic source η in the region S is independent of ξ^- and $\xi = \xi^- + \xi^+$ outside S.

Considering our scheme with the general operator $\mathcal{L}$ in $F = L_2$ we have x, $\operatorname{supp} x \subseteq S^c$, represented as $x = \mathcal{L}^* g + x^+$, $x^+ \in X^+(\Gamma_0)$, where $g \in L_2(S_0)$ are such that $\operatorname{supp} \mathcal{L}^* g \subseteq S^c$, i.e.,

$$(\mathcal{L}\varphi, g) = 0, \qquad \varphi \in C_0^\infty(S),$$

which shows that

$$\mathrm{E}(\mathcal{L}\varphi, \eta)(g, \eta) = 0, \qquad \varphi \in C_0^\infty(S),$$

with $(g, \eta) = (x, \xi^-)$ for the *original* white noise source η on the L_2-space. Hence, we obtain the *new* stochastic source η in (2.7) such that

$$\mathrm{E}(\varphi, \eta)(x, \xi^-) = 0, \qquad \varphi \in C_0^\infty(S),$$

for all x, $\operatorname{supp} x \subseteq S^c$. Thus, this *new* source η in S is independent of ξ^- and $\xi = \xi^- + \xi^+$ outside S. The proof is completed.

It is worth mentioning that considering the model (2.1), (2.2) in $F = L_2$ with some *nonhomogeneous* chaotic source η of the type

$$\mathrm{E}(\varphi, \eta) = 0, \quad \mathrm{E}|(\varphi, \eta)|^2 = \int |\varphi(t)|^2 \sigma^2(t)\, dt, \quad \varphi \in C_0^\infty(S),$$

one can apply $\mathcal{P} = \mathcal{L}^*\mathcal{L}$ in (2.8) with *another* operator $\mathcal{L}$ which results from the original equation (2.1) by multiplication with the function $1/\sigma$.

Let us now consider the forecast

$$\hat{\xi} = \mathrm{E}(\xi/\mathcal{B})$$

in the framework of our scheme (2.1), (2.2) with the arbitrary differential operator $\mathcal{L}$ in the L_2-space and *arbitrary* stochastic source η *with independent values, independent* of the boundary conditions.

Recall that with the required type of boundary conditions (2.2) we have for our test functions $x \in X(S_0) = [C_0^\infty(S_0)]$ in the region S_0 the direct sum representation

$$x = \mathcal{L}^* g + x^+, \qquad g \in L_2(S_0), \qquad x^+ \in X^+(\Gamma_0),$$

with nondegenerate operator $\mathcal{L}^*$. Hence, taking $g = g^- + g^+ \in L_2(S_0)$ with $g^- \in L_2(S)$, $g^+ \in L_2(S^c)$, we obtain the corresponding direct sum representation

$$x = \mathcal{L}^* g^- + \mathcal{L}^* g^+ + x^+,$$

where for every x, $\operatorname{supp} x \subseteq S^c$, we have its component $\mathcal{L}^* g^- \in X^-(\Gamma)$ with $\mathcal{L}^* g^- = 0$ in S, $g^- = 0$ outside S. Thus

$$X(S^c) = \{x \in X(S_0),\ \operatorname{supp} x \subseteq S^c\}$$

is the direct sum

$$X(S^c) = X^-(\Gamma) + X(S^c)^+ \tag{2.10}$$

of the known boundary subspace $X^-(\Gamma) = L_2(S) \cap X(\Gamma)$ on $\Gamma = \partial S$ and

$$X(S^c)^+ = \{\mathcal{L}^* g^+ + x^+,\ g^+ \in L_2(S^c),\ x^+ \in X^+(\Gamma_0)\}.$$

Obviously, the direct sum (2.10) with $X^-(\Gamma) \subseteq X(\Gamma) \subseteq X(S^c)$ gives the corresponding direct sum

$$X(\Gamma) = X^-(\Gamma) + X^+(\Gamma)$$

with

$$X^+(\Gamma) = X(\Gamma) \cap X(S^c)^+. \tag{2.11}$$

Hence, for our test functions in the region $S \subseteq S_0$, we have the direct sum representation

$$X(S) = \mathcal{L}^* L_2(S) + X^+(\Gamma)$$

and, as we encountered many times, by virtue of equation (2.1) in S and the boundary values

$$(x, \xi), \qquad x \in X^+(\Gamma),$$

on the boundary $\Gamma = \partial S$ the random field ξ can be completely identified in the region S.

The stochastic source η in the region S is independent of its part outside S and of the boundary conditions on $\Gamma_0 = \partial S_0$. Hence, η in S is *independent* of the *outside data*

$$\mathcal{A}(S^c)^+ = \{(x, \xi) = (g^+, \eta) + (x^+, \xi^+),\ x \in X(S^c)^+\} \tag{2.12}$$

which can be obtained by measuring the stochastic source η outside the region $S \subseteq S_0$ and the boundary field ξ^+ on the boundary $\Gamma_0 = \partial S_0$. Note, these data

$$\mathcal{B} = \mathcal{A}(S^c)^+$$

represent the total outside data $\mathcal{A}(S^c)$ in the case of $X^-(\Gamma) = 0$ in (2.10), i.e., in the case when there is no $g \in L_2(S)$, $\mathcal{L}^* g = 0$ in S except $g = 0$.

Obviously, regarding the outside data $\mathcal{B} = \mathcal{A}(S^c)^+$ containing the boundary data on the boundary subspace $X^+(\Gamma)$ defined in (2.11), in the general scheme (2.1)′, (2.2)′ the following result holds.

THEOREM. *The forecast $\hat{\xi}$ in the region $S \subseteq S_0$ with the given data $\mathcal{B} = \mathcal{A}(S^c)^+$ outside S can be obtained as the unique solution $u = \hat{\xi} \in \mathbf{W}(S)$ of the homogeneous equation*

$$\mathcal{L}u = 0 \tag{2.13}$$

in S with the boundary conditions

$$(x, u) = (x, \xi), \qquad x \in X^+(\Gamma). \tag{2.14}$$

For illustration, one can consider the Brownian motion $\xi = \xi(t)$, $t \geq t_0$, in the scheme (2.1), (2.2) on the infinite interval $S_0 = (s_0, \infty)$ with $\mathcal{L} = d/dt$ and the Gaussian white noise η *independent* of the boundary condition $\xi(s_0) = \xi^+(s_0)$. In our forecast problem of ξ on the infinite $S = (s, \infty) \subseteq S_0$ the data $\mathcal{B} = \mathcal{A}(S^c)^+$ give the total outside data

$$\mathcal{B} = \mathcal{A}(S^c) = \{\xi(t),\ s_0 \le t \le s\},$$

since there is no $g \in L_2(S)$, $\mathcal{L}^* g = 0$ except $g = 0$. The boundary problem (2.13), (2.14) is

$$\frac{d}{dt} u = 0$$

in $S = (s, \infty)$ with the initial

$$u(s) = \xi(s).$$

Going back to the general case, it is worth mentioning that for the total outside data

$$\mathcal{B} = \mathcal{A}(S^c) = \mathcal{A}(\Gamma)^- \cup \mathcal{A}(S^c)^+$$

with the exclusive boundary values

$$\mathcal{A}(\Gamma)^- = \{(g, \eta),\ g \in L_2(S),\ \mathcal{L}^* g = 0 \text{ in } S\}$$

the forecast equation (2.13) should be replaced by

$$\mathcal{L} u = f \tag{2.13$'$}$$

where

$$f = \mathrm{E}(\eta / \mathcal{A}(\Gamma)^-)$$

in the region S — see (2.1)′, (2.2)′. ■

We now pay particular attention to the following property which we obtain from (2.8), (2.9) or (2.13)–(2.13)′, (2.14): namely, the forecast $\hat{\xi}$ in the region $S \subseteq S_0$ by means of all data $\mathcal{A}(S^c)$ outside S requires only the boundary data $\mathcal{A}(\Gamma)$ on the boundary $\Gamma = \partial S$. This is some kind of *Markov property* of the random field ξ.

In our scheme (2.1), (2.2), we deal with the generalized random field ξ by testing it with the corresponding test functions $x \in X(S_0) = [C_0^\infty(S_0)]$ in the region S_0; for any region $S \subseteq S_0$ it seems natural to introduce the corresponding *Markov property* as conditional independence of

$$\mathcal{A}([S]) = \{(x, \xi) \ \mathrm{supp}\, x \subseteq [S]\}$$

in the region S and

$$\mathcal{A}(S^c) = \{(x, \xi),\ \mathrm{supp}\, x \subseteq S^c\}$$

in the complement S^c outside S, conditioned by the given boundary values

$$\mathcal{A}(\Gamma) = \{(x,\xi),\ \operatorname{supp} x \subseteq \Gamma\}$$

on the boundary $\Gamma = \partial S$. Recall, for the Gaussian random field ξ described by (2.1), (2.2) with the white noise source independent of the boundary conditions we have Markov property according to

$$\mathrm{E}[\xi/\mathcal{A}(S^c)] = \mathrm{E}[\xi/\mathcal{A}(\Gamma)]$$

in any region $S \subseteq S_0$ (see (2.8), (2.9)).

Suppose that this Markov property holds. Then for our forecast problem with the data $\mathcal{B} = \mathcal{A}(S^c)$ outside S, we can apply the equation

$$\mathcal{L}\hat{\xi} = \hat{\eta} = \mathrm{E}[\eta/\mathcal{A}(\Gamma)]$$

in the region S with appropriate boundary conditions

$$(x,\hat{\xi}) = (x,\xi), \qquad x \in X^+(\Gamma),$$

on the boundary $\Gamma = \partial S$ (cf. (2.1)$'$, (2.2)$'$). Let us assume now that we can get some additional data $\mathcal{B}^\varepsilon$ in some neighborhood Γ^ε of the boundary $\Gamma = \partial S$. The question is: How should we adjust the right-hand side $\hat{\eta} = \mathrm{E}[\eta/\mathcal{A}(\Gamma)]$ of the forecast equation?

At first sight, it seems obvious that one should take the corresponding

$$\hat{\eta} = \mathrm{E}[\eta/\mathcal{A}(\Gamma) \cup \mathcal{B}^\varepsilon]$$

with respect to $\mathcal{A}(\Gamma)$ and additional data $\mathcal{B}^\varepsilon$, but this could lead to mistakes, since it could happen that the random field ξ in S conditioned with $\mathcal{A}(\Gamma) \vee \mathcal{B}^\varepsilon$ becomes dependent on some other data in $\mathcal{B} = \mathcal{A}(S^c) \vee \mathcal{B}^\varepsilon$ even though, when conditioned with $\mathcal{A}(\Gamma)$, it was independent on $\mathcal{B} = \mathcal{A}(S^c)$.

For illustration, let us consider the *generalized Brownian motion* ξ on the real line $T = (-\infty, \infty)$ which in our scheme (2.1), (2.2) is represented by the equation

$$\frac{d}{dt}\xi = \eta$$

in $S_0 = T$ with $\mathcal{L} = d/dt$ and the Gaussian white noise η on the $L_2(T)$-space. This *generalized* random function ξ, $\mathrm{E}\xi = 0$, can be described by the complete system of test functions

$$x = \delta_t - \delta_s, \qquad s,t \in T = (-\infty,\infty),$$

in our test functions space $X = \mathcal{L}^* L_2(T) = [C_0^\infty(T)]$ as

$$(x, \xi) \equiv \xi(t) - \xi(s)$$

with

$$\mathrm{E}|\xi(t) - \xi(s)|^2 = |t - s|$$

Taking $S = (0, \infty)$ say, we have no boundary test functions $x = \mathcal{L}^* g$, $\operatorname{supp} x \subseteq \Gamma = \{0\}$, since there is no $g \in L_2(T)$, $\mathcal{L}^* g = 0$ in $T \setminus \Gamma$ with $\mathcal{L}^* = -d/dt$, except $g = 0$. Thus, $\mathcal{A}(\Gamma) = \varnothing$ is empty set, and $\mathcal{A}([S])$ generated by $\xi(t) - \xi(s)$, $s, t \geq 0$, is *independent* of $\mathcal{B} = \mathcal{A}(S^c)$ generated by $\xi(t) - \xi(s)$, $s, t \leq 0$. Suppose that in addition to $\mathcal{A}(S^c)$ we have

$$\mathcal{B}^\varepsilon = \{(\mathcal{L}\varphi, \eta),\ \varphi \in C_0^\infty(\Gamma^\varepsilon)\}$$

generated by the generalized derivative $\eta' = -\mathcal{L}^* \eta$ on the interval $\Gamma^\varepsilon = (-\varepsilon, \varepsilon)$; then $\mathcal{A}([S])$ conditioned with $\mathcal{B}^\varepsilon = \mathcal{A}(\Gamma) \vee \mathcal{B}^\varepsilon$ becomes substantially dependent on $\mathcal{A}(S^c)$. Indeed, in the case of their conditional independence (with respect to $\mathcal{B}^\varepsilon$), we should have conditionally independent $\mathcal{A}([S]) \vee \mathcal{B}^\varepsilon$ and $\mathcal{A}(S^c) \vee \mathcal{B}^\varepsilon$, but this is not so; for example, we have obvious dependence of random variables

$$\eta_1 = [\xi(3\varepsilon) - \xi(\varepsilon)] - [\xi(\varepsilon) - \xi(-\varepsilon)] \in \mathcal{A}([S]) \vee \mathcal{B}^\varepsilon$$

and

$$\eta_2 = [\xi(-3\varepsilon) - \xi(-\varepsilon)] - [\xi(-\varepsilon) - \xi(\varepsilon)] \in \mathcal{A}([S]) \vee \mathcal{B}^\varepsilon$$

which do not depend on $\mathcal{B}^\varepsilon$, and therefore conditioning with $\mathcal{B}^\varepsilon$ does not change η_1 and η_2.

We are to stress here that the Markov property of our general model (2.1), (2.2) implies that the stochastic source η in the region $S \subseteq S_0$ *conditioned* with the boundary data $\mathcal{A}(\Gamma)$ is conditionally independent of its part outside S.

2.2. Global Markov Property

We are going to consider our general scheme (2.1), (2.2) with the differential operator $\mathcal{L}$ in the space $F = L_2(T)$ and the *Markov type* stochastic source η in the region $S_0 \subseteq T$ which is *independent* of the boundary conditions on $\Gamma_0 = \partial S_0$. Here we have in mind the generalized random field

$$\eta = (\varphi, \eta), \qquad \varphi \in C_0^\infty(S_0),$$

meansquare continuous with respect to $\|\varphi\|_{L_2}$ such that in any region $S \subseteq S_0$ it is conditionally independent of its part outside S being conditioned by a σ-algebra $\mathcal{B}(\Gamma)$ of events on the boundary $\Gamma = \partial S$,

$$\mathcal{B}(\Gamma) \subseteq \cap \mathcal{B}(\Gamma^{\varepsilon}) \tag{2.15}$$

where on the right-hand side the intersection

$$\mathcal{B}(\Gamma^{\varepsilon}) = \{(\varphi, \eta), \ \operatorname{supp} \varphi \subseteq \Gamma^{\varepsilon}\}$$

represents a *σ-algebra generated* by η in the neighborhood Γ^{ε} of Γ. In particular, generalized random field η with independent values fits this scheme with the trivial $\mathcal{B}(\Gamma)$. We will show that the random field ξ in (2.1), (2.2) has the following *global Markov property*: for any set $S \subseteq S_0$, the complement S^c and any closed set $\Gamma \supseteq \partial S$, the random field ξ in $S \cup \Gamma$ is conditionally independent of its part in $S^c \cup \Gamma$ being conditioned by its values on Γ.[1]

We recall some properties of the events σ-algebras $\mathcal{A}_1$ and $\mathcal{A}_2$ with *splitting σ-algebra $\mathcal{B}$* which *splits* $\mathcal{A}_1$ and $\mathcal{A}_2$ in a way that $\mathcal{A}_1$ becomes conditionally independent of $\mathcal{A}_2$ with respect to $\mathcal{B}$, i.e., conditional probabilities with respect to $\mathcal{B}$ are such that

$$P(A_1 A_2/\mathcal{B}) = P(A_1/\mathcal{B}) P(A_2/\mathcal{B}) \tag{2.16}$$

for all $A_1 \in \mathcal{A}_1$ and $A_2 \in \mathcal{A}_2$.

With the help of the indicators $\xi_1 = 1_{A_1}$ and $\xi_2 = 1_{A_2}$, this conditional independence can be characterized as

$$\mathrm{E}(\xi_1 \xi_2/\mathcal{B}) = \mathrm{E}(\xi_1/\mathcal{B})(\mathrm{E}(\xi_2/\mathcal{B})$$

and this relation holds for the general random variables ξ_1 and ξ_2 measurable with respect to A_1 and A_2.

The splitting σ-algebra $\mathcal{B}$ splits also the σ-algebras $\mathcal{A}_1 \vee \mathcal{B}$ and $\mathcal{A}_2 \vee \mathcal{B}$ generated correspondingly by the $\mathcal{A}_1$, $\mathcal{A}_2$ and $\mathcal{B}$ itself, for all events $A_1 B_1$ and $A_2 B_2$ are conditionally independent with $A_1 \in \mathcal{A}_1$, $A_2 \in \mathcal{A}_2$, and B_1, $B_2 \in \mathcal{B}$ according to

$$\mathrm{E}(1_{A_1 B_1} \cdot 1_{A_2 B_2}/\mathcal{B}) = 1_{B_1} \cdot 1_{B_2}\, \mathrm{E}(1_{A_1} \cdot 1_{A_2}/\mathcal{B})$$

$$= 1_{B_1} \cdot 1_{B_2} \mathrm{E}(1_{A_1}/\mathcal{B}) \mathrm{E}(1_{A_2}/\mathcal{B}) = \mathrm{E}(1_{A_1 B_1}/\mathcal{B}) \mathrm{E}(1_{A_2 B_2}/\mathcal{B}).$$

[1]For the related material see, for example, Yu.A. Rozanov, Markov Random Fields, Springer, 1981.

For the general random variables ξ_1 and ξ_2 measurable with respect to $\mathcal{A}_1, \mathcal{A}_2 \vee \mathcal{B}$, we have

$$E\{[\xi_1 - E(\xi_1/\mathcal{B}]\xi_2\} = EE\{[\xi_1 - E(\xi_1/\mathcal{B})]\xi_2/\mathcal{B}\} = 0$$

which implies

$$E(\xi_1/\mathcal{B}) = E(\xi_1/A_2 \vee \mathcal{B})$$

and shows that conditional independence of $\mathcal{A}_1$ and $\mathcal{A}_2$ can be characterized as

$$P(A_1/\mathcal{A}_2 \vee \mathcal{B}) = P(A_1/\mathcal{B}), \qquad A_1 \in \mathcal{A}_1. \tag{2.17}$$

It looks first like a paradox that stronger conditioning with $\widetilde{\mathcal{B}} \supseteq \mathcal{B}$ could make $\mathcal{A}_1$ and $\mathcal{A}_2$ substantially dependent on each other (recall our example relating to the generalized Brownian motion). Nevertheles, $\widetilde{\mathcal{B}} \supseteq \mathcal{B}$ splits $\mathcal{A}_1$ and $\mathcal{A}_2$ when it is as

$$\widetilde{\mathcal{B}} = \mathcal{B}_1 \vee \mathcal{B} \vee \mathcal{B}_2 \tag{2.18}$$

generated by $\mathcal{B}$ and $\mathcal{B}_1 \subseteq \mathcal{A}_1$, $\mathcal{B}_2 \subseteq \mathcal{A}_2$; one can verify this for $\widetilde{\mathcal{B}} = \mathcal{B} \vee \mathcal{B}_2$, say, as

$$E(\xi_1/\widetilde{\mathcal{B}}) = E[E(\xi_1/\mathcal{B} \vee \mathcal{A}_2)/\widetilde{\mathcal{B}}]$$

$$= E[E(\xi_1/\mathcal{B})/\widetilde{\mathcal{B}}] = E(\xi_1/\mathcal{B}) = E(\xi_1/\mathcal{A}_2 \vee \widetilde{\mathcal{B}})$$

with $\mathcal{A}_2 \vee \widetilde{\mathcal{B}} = \mathcal{A}_2 \vee \mathcal{B}$.

Of course, the σ-algebras $\mathcal{A}_1$ and $\mathcal{A}_2$ can be interchanged in all relationships considered above.

We should also recall that for splitting σ-algebras $\mathcal{B}_1 \supseteq \mathcal{B}_2 \supseteq \dots$ their intersection

$$\mathcal{B} = \bigcap \mathcal{B}_n \tag{2.19}$$

is also a splitting σ-algebra, because

$$E(\xi_1\xi_2/\mathcal{B}) = \lim E(\xi_1\xi_2/\mathcal{B}_n)$$

$$= \lim[E(\xi_1/\mathcal{B}_n)E(\xi_2/\mathcal{B}_n)] = E(\xi_1/\mathcal{B})E(\xi_2/\mathcal{B})$$

with ξ_1 and ξ_2 measurable with respect to $\mathcal{A}_1$ and $\mathcal{A}_2$. ■

Let us consider the generalized random field

$$\eta = (\varphi, \eta), \qquad \varphi \in C_0^\infty(S_0),$$

with *independent values*, meansquare continuous with respect to $\|\varphi\|_{L_2}$ and well-defined for all $\varphi \in L_2(S_0)$. In order to make terminology convenient, we assume that the region $S_0 \subseteq R^d$ has the zero measure boundary $\Gamma_0 = \partial S_0$ which allows us to introduce formally

$$\eta = (\varphi, \eta), \qquad \varphi \in L_2([S_0]),$$

In any open set $S \subseteq S_0$, the events σ-algebra

$$\mathcal{B}(S) = \{(\varphi, \eta),\ \varphi \in L_2(S)\}$$

generated by (φ, η), $\varphi \in L_2(S)$, is independent of

$$\mathcal{B}(S^c) = \{(\varphi, \eta),\ \varphi \in L_2(S^c)\}$$

outside S, because $\mathcal{B}(S)$ is the limit of the corresponding σ-algebras

$$\{(\varphi, \eta),\ \operatorname{supp}\varphi \subseteq S_{\text{loc}} \subseteq S\}$$

in $S_{\text{loc}} \subseteq S$ which are separated from the complement $S^c = [S_0] \setminus S$.

For a set $\Gamma \subseteq [S_0]$, we have

$$\mathcal{B}(\Gamma) = \mathcal{B}(\Gamma^-) \vee \mathcal{B}(\Gamma^+)$$

generated by $\mathcal{B}(\Gamma^-) \subseteq \mathcal{B}(S)$ in $\Gamma^- = \Gamma \cap S$ and $\mathcal{B}(\Gamma^+) \subseteq \mathcal{B}(S^c)$ in $\Gamma^+ = \Gamma \cap S^c$ according to

$$L_2(\Gamma) = L_2(\Gamma^-) + L_2(\Gamma^+).$$

Hence, the σ-algebra $\mathcal{B}(\Gamma)$ splits the independent $\mathcal{B}(S)$ and $\mathcal{B}(S^c)$, i.e., the σ-algebras $\mathcal{B}(S)$ and $\mathcal{B}(S^c)$ are also *conditionally independent* with respect to $\mathcal{B}(\Gamma)$ for any Γ. ■

Let us now return to our model (2.1), (2.2) with the stochastic source η with independent values in S_0, independent of the boundary conditions on $\Gamma_0 = \partial S_0$.

In this case for any open $S \subseteq S_0$ we have independent

$$\mathcal{A}(S)^- = \{(x, \xi),\ x = \mathcal{L}^* g^-,\ g^- \in L_2(S)\}$$

and

$$\mathcal{A}(S^c)^+ = \{(x, \xi),\ x = \mathcal{L}^* g^+ + x^+,\ g^+ \in L_2(S^c),\ x^+ \in X^+(\Gamma_0)\}.$$

This implies that the σ-algebra

$$\mathcal{A}(\Gamma) = \mathcal{A}(\Gamma)^- \vee \mathcal{A}(\Gamma)^+$$

on the boundary $\Gamma = \partial S$ with

$$\mathcal{A}(\Gamma)^- = \{(x,\xi),\ x \in X^-(\Gamma)\} \subseteq \mathcal{A}(S)^-,$$
$$\mathcal{A}(\Gamma)^+ = \{(x,\xi),\ x \in X^+(\Gamma)\} \subseteq \mathcal{A}(S^c)^+$$

splits the σ-algebras $\mathcal{A}(S)^-$ and $\mathcal{A}(S^c)^+$ as well as the σ-algebras

$$\mathcal{A}([S]) = \{(x,\xi),\ \operatorname{supp} x \subseteq [S]\},$$
$$\mathcal{A}(S^c) = \{(x,\xi),\ \operatorname{supp} x \subseteq S^c\}$$

which are generated as

$$\mathcal{A}([S]) = \mathcal{A}(S)^- \vee \mathcal{A}(\Gamma),$$
$$\mathcal{A}(S^c) = \mathcal{A}(S^c)^+ \vee \mathcal{A}(\Gamma).$$

Moreover, for any closed set $\Gamma \supseteq \partial S$ in $[S_0]$, according to the direct sum representation

$$X(S_0) = \mathcal{L}^* L_2(S) + X(S^c)^+$$

(see (2.10)–(2.12)), with the corresponding

$$X^-(\Gamma) = \{x = \mathcal{L}^* g^-,\ \operatorname{supp} x \subseteq \Gamma\} \subseteq \mathcal{L}^* L_2(S),$$
$$X^+(\Gamma) = \{x = \mathcal{L}^* g^+ + x^+,\ \operatorname{supp} x \subseteq \Gamma\} \subseteq X(S^c)^+$$

and $\mathcal{A}(\Gamma)^-$, $\mathcal{A}(\Gamma)^+$ the same result holds for

$$\mathcal{A}(\Gamma) = \{(x,\xi),\ \operatorname{supp} x \subseteq \Gamma\}. \tag{2.20}$$

Namely, for any closed set Γ, $\partial S \subseteq \Gamma \subseteq [S_0]$, the σ-algebra $\mathcal{A}(\Gamma)$ splits $\mathcal{A}(S)^-$ and $\mathcal{A}(S^c)^+$ as well as the σ-algebras

$$\begin{aligned} \mathcal{A}(S \cup \Gamma) &= \{(x,\xi),\ \operatorname{supp} x \subseteq S \cup \Gamma\}, \\ \mathcal{A}(S^c \cup \Gamma) &= \{(x,\xi),\ \operatorname{supp} x \subseteq S^c \cup \Gamma\} \end{aligned} \tag{2.21}$$

which are generated as

$$\mathcal{A}(S \cup \Gamma) = \mathcal{A}(S)^- \vee \mathcal{A}(\Gamma),$$
$$\mathcal{A}(S^c \cup \Gamma) = \mathcal{A}(S^c)^+ \vee \mathcal{A}(\Gamma).$$

And here we can take *any* set $S \subseteq S_0$, because we actually deal with

$$S \cup \Gamma = \overset{\circ}{S} \cup \Gamma, \qquad S^c \cup \Gamma = \overset{\circ}{S}{}^c \cup \Gamma$$

for the corresponding region $\overset{\circ}{S}$ of all inner points in S, $\partial \overset{\circ}{S} \subseteq \Gamma$.

Let us formulate our result as follows.

THEOREM. *The random field* ξ *in* (2.1), (2.2) *enjoys the global Markov property for any stochastic source* η *with independent values in equation* (2.1), *independent on boundary conditions* (2.2). ■

Similar result holds for the model (2.1), (2.2) with the *Markov type* stochastic source η in S_0 independent of the boundary conditions on $\Gamma_0 = \partial S_0$; we only need to substitute the σ-algebras of events (2.20), (2.21) representing the behaviour of the random field ξ on $S \cup \Gamma$, $S^c \cup \Gamma$, and Γ (for any closed set $\Gamma \supseteq \partial S$) by the corresponding

$$\mathcal{A}(S \cup \Gamma) = \bigcap \mathcal{A}(S \cup \Gamma^\varepsilon),$$
$$\mathcal{A}(S^c \cup \Gamma) = \bigcap \mathcal{A}(S^c \cup \Gamma^\varepsilon), \tag{2.22}$$
$$\mathcal{A}(\Gamma) = \bigcap \mathcal{A}(\Gamma^\varepsilon)$$

where we employ the neighborhoods Γ^ε of Γ in $[S_0]$.

The proof can be done in the same way as above with the starting point changed from the trivial σ-algebra $\mathcal{B}(\Gamma)$ on the boundary $\Gamma = \partial S$ to the boundary σ-algebra (2.15) which, according to the Markov property of η, splits $\mathcal{A}(S)^-$ and $\mathcal{A}(S^c)^+$; it implies that the corresponding

$$\mathcal{A}(\Gamma^\varepsilon) = \mathcal{A}(\Gamma^\varepsilon)^- \vee \mathcal{A}(\Gamma^\varepsilon)^+,$$
$$\mathcal{A}(\Gamma) = \cap \mathcal{A}(\Gamma^\varepsilon)$$

split the σ-algebras

$$\mathcal{A}(S \cup \Gamma) \subseteq \mathcal{A}(S \cup \Gamma^\varepsilon),$$
$$\mathcal{A}(S^c \cup \Gamma) \subseteq \mathcal{A}(S^c \cup \Gamma^\varepsilon).$$

It is worth mentioning that, with multiplication by $w \in C_0^\infty(R^d)$ well-defined for the test functions $x \in X(S_0) = [C_0^\infty(S_0)]$ in $S_0 \subseteq R^d$, we have here the splitting σ-algebras $\mathcal{A}(\Gamma)$ on $\Gamma \supseteq \partial S$ as

$$\mathcal{A}(\Gamma) = \bigcap \{(\varphi, \xi),\ \varphi \in C_0^\infty(\Gamma^\varepsilon)\} \tag{2.23}$$

and correspondingly

$$\mathcal{A}(S \cup \Gamma) = \bigcap \{(\varphi, \xi),\ \varphi \in C_0^\infty(S \cup \Gamma^\varepsilon)\},$$
$$\mathcal{A}(S^c \cup \Gamma) = \bigcap \{(\varphi, \xi),\ \varphi \in C_0^\infty(S^c \cup \Gamma^\varepsilon)\}. \tag{2.24}$$

Recall, that in a case of the stochastic source η being as the random field with independent values we have the splitting σ-algebra on Γ defined as

$$\mathcal{A}(\Gamma) = \{(x,\xi),\ \operatorname{supp} x \subseteq \Gamma\}$$

— see (2.20). And we obtain, particularly, the same σ-algebra $\mathcal{A}(\Gamma)$ in (2.23) when the obvious relationship

$$\bigcap \{(x,\xi),\ \operatorname{supp} x \subseteq \Gamma^\varepsilon\} = \{(x,\xi),\ \operatorname{supp} x \subseteq \Gamma\}$$

for the very random variables considered implies the same relationship for the corresponding σ-algebras; in particular, it holds true when we deal with the nondegenerate Gaussian random field ξ,

$$\mathrm{E}(x,\xi) = 0, \qquad \mathrm{E}|(x,\xi)|^2 \asymp \|x\|_X^2.$$

III.3. A Generalized Approach to Boundary Problems

Our functional approach to boundary problems can be developed in various ways. Similarly to (2.1), (2.2), one can apply, for example, the following scheme associated with some Banach space $W(S)$ of generalized functions in the region S,

$$C_0^\infty(S) \subseteq W(S) \subseteq L_2(S),$$

such that

$$\|\mathcal{L}u\|_{L_2} \leq C\|u\|_W, \qquad u \in W(S), \tag{3.1}$$

for the differential operator $\mathcal{L}$

$$\mathcal{L}u = (\varphi, \mathcal{L}u) \equiv (\mathcal{L}^*\varphi, u), \qquad \varphi \in C_0^\infty(S).$$

By virtue of embedding $W(S) \subseteq L_2(S)$ with $\|u\|_{L_2} \leq C\|u\|_W$, a dual space

$$X(S) = W(S)^*$$

contains all $x \in C_0^\infty(S)$, because

$$\|(x,u)\| \leq \|x\|_{L_2}\|u\|_{L_2} \leq C\|u\|_W.$$

Moreover

$$X(S) = [C_0^\infty(S)]$$

in the case $X(S)^* = W(S)$, since $(x,u) = 0$, $x \in C_0^\infty(S)$, implies $u = 0$ for Schwartz distributions $u \in W(S)$ in the region S.

One should treat $x \in X(S)$ as test functions which can be applied to all generalized random functions

$$u = (x, u), \qquad x \in C_0^\infty(S),$$

which are meansquare continuous with respect to $\|x\|_X$ and well-defined as

$$u = (x, u), \qquad x \in X(S) = [C_0^\infty(S)].$$

Dealing with this collection $u \in \mathbf{W}(S)$ of generalized random functions, one can consider the generalized differential equation

$$\mathcal{L}u = f \tag{3.2}$$

in the region S with boundary conditions of the type

$$(x, u) = (x, u^+), \qquad x \in X^+(\Gamma), \tag{3.3}$$

which are set by means of some boundary test functions x, $\operatorname{supp} x \subseteq \Gamma$, on the boundary $\Gamma = \partial S$, i.e., such that

$$(x, \varphi) \equiv (\varphi, x) = 0, \qquad \varphi \in C_0^\infty(S).$$

— cf. (2.11), Chapter 2.

The test functions $x = \mathcal{L}^*\varphi$, $\varphi \in C_0^\infty(S)$, which are employed in equation (3.2), with application to the deterministic functions $u \in W(S)$, $\mathcal{L}u = f \in L_2(S)$, are such that

$$\|\mathcal{L}^*\varphi\|_X = \sup_{\|u\|_W \leq 1} |(\mathcal{L}^*\varphi, u)| = \sup_{\|u\|_W \leq 1} \|(\varphi, \mathcal{L}u)\|$$

$$\leq \sup_{\|u\|_W \leq 1} \|\varphi\|_{L_2} \|\mathcal{L}u\|_{L_2} \leq C\|\varphi\|_{L_2}, \qquad \varphi \in C_0^\infty(S).$$

And in the case when equation (3.2) is solvable for all $f \in L_2(S)$, we have

$$\|\mathcal{L}^*\varphi\|_X \asymp \|\varphi\|_{L_2}, \qquad \varphi \in C_0^\infty(S). \tag{3.5}$$

Indeed, as it follows from the uniform boundness principle,

$$\|\varphi\|_{L_2} \leq C$$

on the set of all φ: $\|\mathcal{L}^*\varphi\|_X \leq 1$ where

$$|(\varphi, f)| = |(\mathcal{L}^*\varphi, u)| \leq \|u\|_W$$

for all $f = \mathcal{L}u \in L_2(S)$.

Considered on $C_0^\infty(S) \subseteq L_2(S)$, the bounded operator $\mathcal{L}^*$: $C_0^\infty(S) \to X$ can be extended to

$$\mathcal{L}^*\colon L_2(S) \to X(S).$$

Thus,

$$\mathcal{L}^* L_2(S) \subseteq X,$$

where with $x = \mathcal{L}^* g$, $g \in L_2(S)$ we have

$$(x, u) = \lim_{\varphi \to g} (\mathcal{L}^* \varphi, u) = \lim_{\varphi \to g} (\varphi, \mathcal{L} u) = (g, \mathcal{L} u), \qquad u \in W(S).$$

Dealing with the Banach space $W(S)$ such that $W(S) = X(S)^*$, let us assume that for any $f \in L_2(S)$ *there is a unique solution* $u \in W(S)$ *of the deterministic boundary problem* (3.2), (3.3) *with zero-boundary conditions.* Then our test functions space $X(S) = [C_0^\infty(S)]$ has the structure of a direct sum

$$X(S) = \mathcal{L}^* L_2(S) + X^+(\Gamma) \tag{3.6}$$

(cf. (2.17), (2.18), Chapter 2). Let us recall that this structure arises if and only if there is a unique linear continuous functional $u = (x, u)$, $x \in X(S)$, which is arbitrary on the subspace $X^-(S) = \mathcal{L}^* L_2(S)$ and zero on the subspace $X^+(\Gamma) \supseteq X(S)$. This implies the following result.

THEOREM. *There is a unique solution* $u \in \mathbf{W}(S)$ *of the stochastic boundary problem* (3.2), (3.3) *for any generalized random source* $f = (\varphi, f)$, $\varphi \in C_0^\infty(S)$, *meansquare continuous with respect to* $\|\varphi\|_{L_2}$, *and any random boundary sample* $u^+ \in \mathbf{W}(S)$. ■

One might be interested in the Forecast problem or Markov property of the generalized random field $\xi = u \in \mathbf{W}(S)$ in the framework of the Probability model (3.2), (3.3).

Let us consider this model in the form

$$\mathcal{L}\xi = \eta$$

in the region $S = S_0$ with the boundary conditions

$$(x, \xi) = (x, \xi^+), \qquad x \in X^+(\Gamma_0),$$

on the boundary $\Gamma_0 = \partial S_0$.

According to (3.6) with $S = S_0$ and the nondegenerate operator $\mathcal{L}^*$ on $L_2(S_0) = L_2(S) + L_2(S^c)$, for any $S \subseteq S_0$ we have the corresponding direct sum representation

$$X(S_0) = \mathcal{L}^* L_2(S) + \mathcal{L}^* L_2(S^c) + X^+(\Gamma_0)$$

and for the test functions subspace

$$X(S) = [C_0^\infty(S)] \subseteq X(S_0)$$

with $\mathcal{L}^* L_2(S) = [\mathcal{L}^* C_0^\infty(S)] \subseteq X(S)$ in the region $S \subseteq S_0$, we have the direct sum

$$X(S) = \mathcal{L}^* L_2(S) + X^+(\Gamma)$$

where

$$X^+(\Gamma) = X(S) \cap [\mathcal{L}^* L_2(S^c) + X^+(\Gamma_0)]. \tag{3.7}$$

Regarding ξ in S as the generalized random field

$$\xi = (x, \xi), \qquad x \in X(S) = [C_0^\infty(S)],$$

one can treat $x \in X^+(\Gamma)$ as *boundary test functions*, $\operatorname{supp} x \subseteq \Gamma$. The boundary test functions space $X(\Gamma)$ of all $x \in X(S)$, $\operatorname{supp} x \subseteq \Gamma$, can be decomposed into the direct sum

$$X(\Gamma) = X^-(\Gamma) + X^+(\Gamma)$$

with

$$X^-(\Gamma) = X(\Gamma) \cap \mathcal{L}^* L_2(S) \tag{3.8}$$

and the subspace $X(S^c)$ of all test functions $x \in X(S_0)$, $\operatorname{supp} x \subseteq S^c$, i.e., such that

$$(x, \varphi) = 0, \qquad \varphi \in C_0^\infty(S),$$

can be decomposed into the direct sum

$$X(S^c) = X^-(\Gamma) + \mathcal{L}^* L_2(S^c) + X^+(\Gamma_0)$$

(cf. (2.10)–(2.12)). As we know, this structure of our test functions (with $X^+(\Gamma)$, $X^-(\Gamma)$ defined in (3.7), (3.8)) leads to the following result for the random field described in the region S_0 as $\xi = u \in \mathbf{W}(S)$ by equation (3.2) with the generalized stochatic source $\eta = f$ with independent values in S_0, independent of the stochastic boundary conditions (3.3) on the boundary $\Gamma_0 = \partial S_0$.

THEOREM. *The generalized random field ξ enjoys the global Markov Property. The Forecast $\hat{\xi} = \mathrm{E}(\xi/\mathcal{A}(S^c))$ of ξ in any region $S \subseteq S_0$ by means of all data outside S can be given as the unique solution $\hat{\xi} = u \in \mathbf{W}(S)$ of the boundary problem of the type* (3.2), (3.3) *with*

$$f = \mathrm{E}(\eta/\mathcal{A}^-(\Gamma)), \qquad \mathcal{A}^-(\Gamma) = \{(x,\xi),\ x \in X^-(\Gamma)\}$$

and

$$u^+ = (x, u^+) = (x, \xi), \qquad x \in X^+(\Gamma).$$

Let us recall that here

$$f = 0$$

if there is no $g \in L_2(S)$ satisfying condition

$$\mathcal{L}^* g = (\varphi, \mathcal{L}^* g) = (\mathcal{L}\varphi, g) = 0, \qquad \varphi \in C_0^\infty(S),$$

except $g = 0$, which actually happens for a number of various differential operators in unbounded regions.

Chapter IV

Gaussian Random Fields

IV.1. Some General Properties

1.1. Gaussian random variables and σ-algebras of events

Recall, a family $\{\xi\}$ of real valued random variables is called *Gaussian* if the joint probability distribution of these (taken in finite number) random variables is Gaussian. The Gaussian probability distribution in R^n of random variables $(\xi_1, \dots, \xi_n)$ has the characteristic function

$$\mathrm{E}\exp\left\{i\sum_{k=1}^{n}\lambda_k\xi_k\right\}$$

$$= \exp\left\{i\sum_{k=1}^{n}a_k\lambda_k - \frac{1}{2}\sum_{k,j=1}^{n}b_{k,j}\lambda_k\lambda_j\right\}, \qquad (\lambda_1, \dots, \lambda_n) \in R^n,$$

with parameters

$$a_k = \mathrm{E}\xi_k, \quad b_{kj} = \mathrm{E}(\xi_k - a_k)(\xi_j - a_j), \qquad k, j = 1, \dots, n;$$

for $\xi_1, \dots, \xi_n$ with zero means and nondegenerate correlation matrix $B = \{b_{kj}\}$ in R^n, there exists the probability density

$$\frac{1}{(2\pi)^{1/2}\det B^{1/2}}\exp\left\{-\frac{1}{2}(B^{-1}x, x)\right\}, \qquad x = (x_1, \dots, x_n) \in R^n. \tag{1.1}$$

And in the Hilbert space $\mathbf{H}$ of random variables ξ, $\|\xi\| = (\mathrm{E}|\xi|^2)^{1/2}$, a (real) linear closure of the arbitrary family $\{\xi\}$ of Gaussian random variables provides us with Gaussian variables as well.

The integral of the Gaussian function. Let us consider the Gaussian function $\xi(t)$, $t \in T$, regarded as a function in the Hilbert space $\mathbf{H}$. Recall that a piecewise constant function $\xi(t)$ taking its values ξ_k on measurable sets $\Delta_k \subseteq T$

$$\xi(t) = \xi_k, \quad k \in \Delta_k, \quad k = 1, 2, \dots, \qquad \Big(\bigcup_k \Delta_k = T\Big)$$

is integrable if the function $\|\xi(t)\|$, $t \in T$, is integrable; then

$$\int_T \xi(t)\, dt = \sum_k \xi_k |\Delta_k|.$$

A general function $\xi(t)$ is integrable if there exists a sequence of integrable piecewise constant functions $\xi_n(t)$ such that

$$\lim \int_T \|\xi(t) - \xi_n(t)\|\, dt = 0.$$

Then

$$\int_T \xi(t)\, dt = \lim \int_T \xi_n(t)\, dt.$$

It is clear that the approximating functions $\xi_n(t)$ can always be taken with values in a linear closure of the random variables $\xi(t)$, $t \in T$ (we denote it by H), since, for the projections $\hat{\xi}(t)$ of the random variables $\xi_n(t)$ onto H, we have

$$\|\xi(t) - \hat{\xi}_n(t)\| \leq \|\xi(t) - \xi_n(t)\|.$$

Thus

$$\int_T \xi(t)\, dt \in H,$$

where, for Gaussian random variables $\xi(t)$, $t \in T$, we have H formed by Gaussian variables.

The function $\xi(t)$, $t \in T$, in the Hilbert space $\mathbf{H}$ is integrable if and only if it has a modification on the probability space Ω such that

$$\xi(t) = \xi(\omega, t), \qquad (\omega, t) \in \Omega \times T,$$

as scalar function of the pair of variables (ω, t) is measurable and

$$\int_T \|\xi(t)\|\, dt \leq \infty,$$

since it obviously holds true for the approximating piecewise constant functions $\xi_n(t) = \xi_n(\omega, t)$, $(\omega, t) \in \Omega \times T$. And the integral $\int \xi(t)\,dt$, regarded as the random variable in $\mathbf{H}$, is such that

$$\int_T \xi(t)\,dt = \int_T \xi(\omega, t)\,dt, \qquad \omega \in \Omega, \tag{1.2}$$

where the right-hand side is the integral of realizations $\xi_\omega = \xi(\omega, t)$, $t \in T$, which are all integrable for a proper modification of the random function $\xi = \xi(t)$, $t \in T$, with

$$\mathrm{E} \int |\xi(t)|\,dt = \int \mathrm{E}|\xi(t)|\,dt \le \int \|\xi(t)\|\,dt.$$

Gausssian σ-algebra of events. Considering a σ-algebra $\mathcal{A}$ of events generated by Gaussian random variables $\{\xi\}$, we can pass to their linear closure H as a Gaussian space in our Hilbert space $\mathbf{H}$ of random variables ξ, $\|\xi\| = \mathrm{E}(|\xi|^2)^{1/2}$. The following property holds true: *An intersection $\cap \mathcal{A}$ of Gaussian σ-algebras is generated by an intersection $\cap H$ of the corresponding Gaussian spaces.*

To prove it, we can consider just a sequence of H_n, $n = 0, 1, \ldots$, assuming that $H_n = \cap_{k \le n} H_k$ formed with zero mean Gaussian variables. We employ the orthogonal complements

$$H_n^\perp = H_n \ominus H_{n-1}, \qquad n = 1, 2, \ldots;$$

these orthogonal spaces $H_n^\perp$ of the zero mean Gaussian variables are independent. Let $\mathcal{A}_n^\perp$ be the corresponding independent σ-algebras (generated by $H_n^\perp$). We have

$$\mathcal{A}_{n-1} = \{\mathcal{A}_n \cup \mathcal{A}_n^\perp\}$$

generated by the indicated union of the independent σ-algebras $\mathcal{A}_n$ and $\mathcal{A}_n^\perp$. And we have

$$\mathcal{A}_0 = \left\{ \bigcup_n \mathcal{A}_n \right\} = \{\mathcal{A} \cup \mathcal{B}\},$$

where $\mathcal{A}$ is generated by the Gaussian space

$$H = \bigcap H_n$$

and $\mathcal{B}$ is generated by all $H_n^\perp$,

$$\mathcal{B} = \left\{ \bigcup_n \mathcal{A}_n^\perp \right\}.$$

Let us consider the space $\mathbf{H}(\mathcal{A}) \in \mathbf{H}$ of all random variables measurable with respect to the σ-algebra $\mathcal{A}$, and its orthogonal complement $\mathbf{H}(\mathcal{A})^{\perp}$ in the corresponding space $\mathbf{H}(\mathcal{A}_0)$ of all random variables measurable with respect to the σ-algebra $\mathcal{A}_0 = \{\cup_n \mathcal{A}_n\}$. Taking a complete system in $\mathbf{H}(\mathcal{A}_0)$ formed by products $1_A \cdot 1_B$ of all indicators 1_A, 1_B with $A \in \mathcal{A}$ and $B \in \mathcal{L}$, from a decomposition

$$1_A \cdot 1_B = (\mathrm{E}1_B)1_A - 1_A(1_B - \mathrm{E}1_B)$$

with $1_A \in \mathbf{H}(\mathcal{A})$ and

$$1_A(1_B - \mathrm{E}1_B) \perp \mathbf{H}(\mathcal{A}),$$

we see that all products $1_A(1_B - \mathrm{E}1_B)$ constitute a complete system in $\mathbf{H}(\mathcal{A})^{\perp}$. Any random variable $\xi \in \mathbf{H}(\mathcal{A}_0)$ which is measurable with respect to all σ-algebras $\mathcal{A}_n$, being independent on all $\mathcal{A}_n^{\perp}$ and $\mathcal{B} = \{\cup_n \mathcal{A}_n^{\perp}\}$, is such that $\mathrm{E}(\xi 1_A)(1_B - \mathrm{E}1_B) = 0$, i.e., it is orthogonal to all $1_A(1_B - \mathrm{E}1_B)$,

$$\xi \perp \mathbf{H}(\mathcal{A})^{\perp},$$

and therefore $\xi \in \mathbf{H}(\mathcal{A})$ is measurable with respect to the σ-algebra $\mathcal{A}$. This ends the proof.

1.2. Polynomials of Gaussian random variables

Let $\{\xi\}$ be an arbitrary family of Gaussian random variables and let $H^p = H^p(\xi)$ be meansquare closure of all polynomials of degree not higher than p of the random variables ξ. Including the random variable $\xi = 1$ into this family, we can say that H^p is generated by the products $\xi_1 \cdots \xi_k$, $k \le p$, of all random variables in $\{\xi\}$. In particular, $H = H^1$ is generated by the family $\{\xi\}$ itself. Obviously, H^p does not change if as the initial family we take $\{\eta\}$ consisting of all random variables $\eta \in H$,

$$H^p(\eta) = H^p(\xi).$$

Conditional expectations. Let us consider an arbitrary family $\{\eta\} \subseteq H(\xi)$, an arbitrary random variable $\phi(\xi) \in H^p(\xi)$ and its conditional expectation with respect to $\{\eta\}$. It turns out that

$$\mathrm{E}[\phi(\xi)/\{\eta\}] \in H^p(\eta). \tag{1.3}$$

Let us prove it. If $p = 1$, for the Gausssian random variable $\phi_1(\xi) = \xi \in H^1(\xi)$ we obtain

$$\mathrm{E}[\phi_1(\xi)/\{\eta\}] = \hat{\xi}$$

as the projection $\hat{\xi}$ of the random variable ξ onto the subspace $H^1(\eta)$ because the Gaussian random variable $\xi - \hat{\xi}$, which is orthogonal to $H^1(\eta)$, does not depend on η and

$$\mathrm{E}[\xi - \hat{\xi}/\{\eta\}] = \mathrm{E}(\xi - \hat{\xi}) = 0.$$

Assuming that (1.3) is valid for all $p < q$ and defining the product $\phi_q = \xi_1 \cdots \xi_q$ for $p = q$, we take the projections $\hat{\xi}_1, \ldots, \hat{\xi}_q$ and set the product

$$\psi_q(\xi) = (\xi_1 - \hat{\xi}_1) \cdots (\xi_q - \hat{\xi}_q).$$

The random variable $\psi_q(\xi) \in H^q(\xi)$ does not depend on $\{\eta\}$ and

$$\mathrm{E}[\psi_q(\xi)/\{\eta\}] = \mathrm{E}\psi_q(\xi).$$

The difference of the polynmomials $\phi_q(\xi) - \psi_q(\xi)$ can be represented by the linear combination of products of the form

$$(\xi_{k_1} \cdots \xi_{k_p})(\hat{\xi}_{k_{p+1}} \cdots \hat{\xi}_{k_q}), \qquad p < q,$$

for which

$$\mathrm{E}[(\xi_{k_1} \cdots \xi_{k_p})(\hat{\xi}_{k_{p+1}} \cdots \hat{\xi}_{k_q})/\{\eta\}] = \mathrm{E}[(\xi_{k_1} \cdots \xi_{k_p})/\{\eta\}]\cdot(\hat{\xi}_{k_{p+1}} \cdots \hat{\xi}_{k_q}),$$

where, according to our assumption that (1.3) holds for $p < q$, we have

$$\mathrm{E}[(\xi_{k_1} \cdots \xi_{k_p})/\{\eta\}] \in H^p(\eta)$$

and together with $\xi_{k_{p+1}}, \ldots, \xi_{k_q} \in H^1(\eta)$, this yields

$$\mathrm{E}[(\xi_{k_1} \cdots \xi_{k_p})/\{\eta\}] \cdot (\hat{\xi}_{k_{p+1}} \cdots \hat{\xi}_{k_q}) \in H^q(\eta).$$

Orthogonal expansion over polynomials. For an arbitrary Gaussian family $\{\xi\}$, all polynomials constitute a complete system in the space

$$\mathbf{H}(\xi) = \mathbf{H}(\mathcal{A})$$

of all random variables which are measurable with respect to the σ-algebra $\mathcal{A}$ generated by $\{\xi\}$.

Turning to the subspaces $H^p = H^p(\xi)$ generated by all polynomials $\phi(\xi)$ of degree not higher than p, we can, by making use of the orthogonalization procedure, pass to the subspaces

$$H_p(\xi) = H^p(\xi) \ominus H^{p-1}(\xi), \qquad p = 1, 2, \ldots, \tag{1.4}$$

including in this sequence $H_0(\xi) = H^0(\xi)$ formed by all constants. The elements

$$\phi(\xi) \in H_p(\xi)$$

obtained in this way are called the *Hermite polynomials* of the Gaussian variables in $\{\xi\}$. Clearly, the orthogonal expansions

$$H^p(\xi) = \sum_{k=0}^{p} \oplus H_k(\xi), \qquad \mathbf{H}(\xi) = \sum_{k=0}^{\infty} \oplus H_k(\xi) \tag{1.5}$$

hold.

The orthogonalization of polynomials $\phi(\xi)$ requires calculation of the expectations of products of the random variables in $\{\xi\}$; note that for the product $\xi_1 \cdots \xi_n$ of any Gaussian random variables with zero mean we have

$$\mathrm{E}(\xi_1 \cdots \xi_n) = \sum \prod b_{kj}, \qquad b_{kj} = \mathrm{E}\xi_k \xi_j,$$

where the sum is taken over all decompositions of the set $(1, \ldots, n)$ into pairs (k, j) of its elements and the product is taken over all such pairs in the corresponding decomposition.

Let us consider two arbitrary families $\{\xi\}$ and $\{\eta\}$ of random variables with the only condition that their joint probability distributions are Gaussian. One would think that for different $H_p(\xi)$ and $H_q(\eta)$ there is no reason to expect any special relations, but in fact *the following orthogonality relation holds: for* $p \neq q$

$$H_p(\xi) \perp H_q(\eta). \tag{1.6}$$

Let us prove it for $p < q$. We consider an arbitrary random variable $\phi = \phi(\xi) \in H^p(\xi)$ and its conditional expectation

$$\hat{\phi} = \mathrm{E}[\phi/\{\eta\}] \in H^p(\eta)$$

(see (1.3)). As known, $\hat{\phi}$ is the projection of the variablye ϕ onto the corresponding space $\mathbf{H}(\eta) = H^p(\eta) \oplus \sum_{k>p} H_k(\eta)$ and, in particular,

$\phi - \hat{\phi} \perp H_q(\eta)$, $q > p$, which together with the orthogonality $H^p(\eta) \perp H_q(\eta)$ for $\hat{\phi} \in H^p(\eta)$ yields $\phi = (\phi - \hat{\phi}) + \hat{\phi} \perp H_q(\eta)$.

According to the properties of the Hermite polynomials of the Gaussian random variables $\{\xi\}$ established above, for any family

$$\{\eta\} \subseteq H(\xi) = H_1(\xi)$$

the subspace $\mathbf{H}(\eta) \subseteq \mathbf{H}(\xi)$ turns out to be invariant under operation of projection onto any subspace $H_p(\xi)$, namely

$$\mathbf{H}(\eta) \to H_p(\eta)$$

since in the orthogonal expansion $\mathbf{H}(\eta) = \sum_{k=0}^{\infty} \oplus H_k(\eta)$ all $H_k(\eta)$, $k \neq p$, are orthogonal to

$$H_p(\xi) \supseteq H_p(\eta).$$

We note that in the classical analysis, the *Hermite polynomials* are defined as the polynomials $\phi_p(x)$ of the real variable x which are orthogonal for different indices $p = 0, 1, \ldots$ with the weight-function

$$p(x) = \frac{1}{\sqrt{2\pi}} e^{-x^2/2},$$

$$\int_{-\infty}^{\infty} \phi_p(x)\phi_q(x)p(x)\,dx = \begin{cases} 0, & p \neq q, \\ 1, & p = q. \end{cases}$$

For any Gaussian family $\{\xi\}$, by taking in $H = H_1(\xi)$ a complete orthonormal system $\{\eta\}$, we can obtain the complete orthonormal system in the whole space $\mathbf{H}(\xi) = \sum_{k=0}^{\infty} H_k(\xi)$ by forming all random variables

$$\phi_{k_1}(\eta_1) \cdots \phi_{k_n}(\eta_n)$$

with various $\eta_1, \ldots, \eta_n$ from the system $\{\eta\}$.

1.3. Comparison theorem for quadratic forms of Gaussian random variables

We consider certain relationships for quadratic formms

$$\phi(\xi) = \sum_{j,k=1}^{n} c_{jk}\xi_j\xi_k$$

of a Gaussian variables with different probability distributions $\mathbf{P}$ and $\mathbf{P}_0$. (It actually concerns two Gaussian families $\{\xi\}$ and $\{\xi^0\}$ with distributions $\mathbf{P}$ and $\mathbf{P}_0$ and given bijective correspondence

$$\xi \leftrightarrow \xi^0$$

between their random variables by comparing the variances of the corresponding quadratic formes

$$\phi(\xi) = \sum_{j,k=1}^{n} c_{jk}\xi_j\xi_k, \qquad \phi(\xi^0) = \sum_{j,k=1}^{n} c_{jk}\xi_j^0\xi_k^0).$$

Let us consider Gaussian variables $\xi_1, \ldots, \xi_n$ with zero mean and the correlation

$$B_{jk} = \mathrm{E}\xi_j\xi_k, \qquad k = 1, \ldots, n.$$

For symmetric forms

$$\phi(\xi) - \mathrm{E}\phi(\xi) \in H_2(\xi)$$

with coefficients $c_{jk} = c_{kj}$, we obtain

$$\mathrm{E}|\phi(\xi) - \mathrm{E}\phi(\xi)|^2 = 2\sum_{j,k}\sum_{l,m} c_{jk}c_{lm}B_{jl}B_{km}.$$

Using the probability space Ω with arbitrary Gaussian distribution $\mathbf{P}$, given on the σ-algebra $\mathcal{A}$ generated by the Gaussian family $\{\xi\}$, and taking the product $\Omega \times \Omega$ with the probability measure $\mathbf{P} \times \mathbf{P}$, we turn to the symmetric bilinear form

$$\phi(\omega, \omega') = \sum_{j,k} c_{jk}\xi_j(\omega)\xi_k(\omega')$$

of $(\omega, \omega') \in \Omega \times \Omega$. It can be easily verified that for such a form we have

$$\iint |\phi(\omega, \omega')|^2\, \mathbf{P}(d\omega)\, \mathbf{P}(d\omega')$$

$$= \sum_{j,k}\sum_{l,m} c_{jk}c_{lm}B_{jl}B_{km} = \frac{1}{2}\mathrm{E}|\phi(\xi) - \mathrm{E}\phi(\xi)|^2. \tag{1.7}$$

Now, let us consider different $\mathbf{P}$ and $\mathbf{P}_0$. Let us assume that for all linear forms of $\{\xi\}$ we have

$$\mathrm{E}\Big|\sum_k c_k\xi_k\Big|^2 \le C\mathrm{E}_0\Big|\sum_k c_k\xi_k\Big|^2 \tag{1.8}$$

where the expectations are taken with respect to $\mathbf{P}$ and $\mathbf{P}_0$. Then, *for all quadratic forms* $\phi(\xi)$ *the inequality*

$$\mathrm{E}|\phi(\xi) - \mathrm{E}\phi(\xi)|^2 \leq C^2 \mathrm{E}_0 |\phi(\xi) - \mathrm{E}_0\phi(\xi)|^2 \tag{1.9}$$

holds.

Let us prove it. Obviously, by virtue of (1.8) we have the relations

$$\iint |\phi(\omega, \omega')|^2 \, \mathbf{P}(d\omega) \times \mathbf{P}(d\omega')$$

$$= \int \mathrm{E} \left| \sum_j \left[\sum_k c_{jk} \xi_k(\omega') \right] \xi_j \right|^2 \mathbf{P}(d\omega')$$

$$\leq C \int \mathrm{E}_0 \left| \sum_j \left[\sum_k c_{jk} \xi_k(\omega') \right] \xi_j \right|^2 \mathbf{P}(d\omega')$$

$$= C \iint |\phi(\omega, \omega')|^2 \, \mathbf{P}_0(d\omega) \times \mathbf{P}(d\omega')$$

which can be extended to the inequality

$$\iint |\phi(\omega, \omega')|^2 \, \mathbf{P}(d\omega) \times \mathbf{P}(d\omega') \leq C^2 \iint |\phi(\omega, \omega')|^2 \, \mathbf{P}_0(d\omega) \times \mathbf{P}_0(d\omega')$$

which, according to (1.7), leads directly to (1.9) with the same C as in (1.8).

1.4. The likelihood ratio

Let us consider a Gaussian family $\{\xi\}$ with two different probability distributions $\mathbf{P}$ and $\mathbf{P}_0$. Having some probability space Ω with random variables $\xi = \xi(\omega)$, $\omega \in \Omega$, we can turn to the corresponding probability measures $\mathbf{P}$ and $\mathbf{P}_0$ on the σ-algebra $\mathcal{A}$ generated by $\{\xi\}$. A question concerns the density

$$p(\omega) = \frac{\mathbf{P}(d\omega)}{\mathbf{P}_0(d\omega)}$$

regarded as a function of random variables in $\{\xi\}$ labelled by the index $u \in U$ as

$$\xi(u) = \xi(\omega, u), \qquad \omega \in \Omega, \qquad u \in U.$$

Recall, the Gaussian distributions $\mathbf{P}$ and $\mathbf{P}_0$ are determined by their *mean*

$$A(u) = \mathrm{E}\xi(u), \qquad A_0(u) = \mathrm{E}_0\xi(u)$$

and *correlation*

$$B(u,v) = \mathrm{E}[\xi(u) - A(u)][\xi(v) - A(v)],$$

$$B_0(u,v) = \mathrm{E}_0[\xi(u) - A_0(u)][\xi(v) - A_0(v)]$$

treated as functions of $u, v \in U$.

Let us first consider $\mathbf{P}$ and $\mathbf{P}_0$ with equal correlation $B = B_0$ when the difference in the disrtributions $\mathbf{P}$ and $\mathbf{P}_0$ exhibits itself in the difference between A and A_0. We take $A_0 = 0$. Clearly, we can always pass to this case by using the transformation

$$\xi(u) \to \xi(u) - A_0(u), \qquad u \in U,$$

under which

$$A_0(u) \to 0, \qquad A(u) \to a(u) = A(u) - A_0(u), \qquad u \in U.$$

Considering the Gaussian family $\{\xi\}$ relative to the distribution $\mathbf{P}_0$, we apply the corresponding Gaussian space $H = H_1(\xi)$. The random variables $\xi = \xi(u)$, $u \in U$, have the mean values $a(u) = \mathrm{E}\xi(u)$, $u \in U$, with respect to the distribution $\mathbf{P}$. It turns out that for the density $p(\omega) = \mathbf{P}(d\omega)/\mathbf{P}_0(d\omega)$ *the formula*

$$p = \sigma^2 e^{\eta} \tag{1.10}$$

holds where $\eta = \phi(\omega) \in H_1(\xi)$ *is uniquely determined from the equation*

$$a(u) = \int \xi(\omega, u)\phi(\omega)\, \mathbf{P}_0(d\omega), \qquad u \in U, \tag{1.11}$$

and the representation (1.11) *of the mean* $a(u)$, $u \in U$, *with some* $\eta = \phi \in H$ *is a necessary and sufficient condition for existence of the density* p; *the constant* σ^2 *in* (1.10) *is determined by the normalization*

$$\int p(\omega)\, \mathbf{P}_0(d\omega) = 1$$

(*one can verify that* $\sigma^2 = \exp\{-\frac{1}{2}\mathrm{E}_0\eta^2\}$).

Clearly, the representation (1.11) and the random variable $\eta \in H = H_1(\xi)$ do not depend on the labelling $\xi = \xi(u)$ by the variable

$u \in U$ and (1.11) expresses the fact that the mean $a(u) = \mathrm{E}\xi(u)$, $u \in U$, determines a linear continuous functional on $\xi \in H$ which on the linear form $\sum c_k \xi(u_k)$ takes the value $\sum c_k a(u_k)$. Then, (1.11) is, in fact, the Riesz representattion of this linear functional on the complete system of elements $\xi(u)$, $u \in U$.

The easiest way to obtain (1.10) and (1.11) is to turn first to a finite number of orthogonal variables $\{\xi(u),\ u = 1, \ldots, n\}$ in $H = H_1(\xi)$ and to the σ-algebra $\mathcal{A}_n$ generated by them. On this σ-algebra, the corresponding density $p_n(\omega) = \mathbf{P}(d\omega)/\mathbf{P}_0(d\omega)$, by the general formula (1.1) for the Gaussian density in R^n, takes the form

$$p_n = \exp\left\{\sum_{u=1}^{n} a(u)\xi(u) - \frac{1}{2}\sum_{u=1}^{n} a^2(u)\right\},$$

where

$$a(u) = \mathrm{E}\xi(u) = \int \xi(\omega, u)\phi_n(\omega)\,\mathbf{P}_0(d\omega)$$

are the coefficients of the orthogonal expansion for the random variable $\eta_n = \sum_{u=1}^{n} a(u)\xi(u) = \phi_n(\omega)$. It is easy to see that for widening finite sets $\{\xi(u),\ u = 1, \ldots, n\}$, the existence of the limit

$$p = \lim p_n$$

is equivalent to the existence of the limit $\eta = \lim \eta_n$ and conditions for this limit existence can be expressed in the form (1.11) which leads us to the limit density (1.10). ■

Let us note that the probability distribution $\mathbf{P}$ of $\{\xi\}$ can be absolutely continuous with respect to $\mathbf{P}_0$ only under the following equivalence condition

$$\mathrm{E}\left|\sum_k c_k \xi_k\right|^2 \asymp \mathrm{E}_0\left|\sum_k c_k \xi_k\right|^2 \tag{1.12}$$

for all linear forms $\sum_k c_k \xi_k$ with ξ_k in $\{\xi\}$; obviously, if this condition is violated, $\mathbf{P}$ and $\mathbf{P}_0$ are mutually singular. Along with (1.12), we have another equivalence condition

$$\iint |\phi(\omega, \omega')|^2\,\mathbf{P}(d\omega)\,\mathbf{P}_0(d\omega') \asymp \mathrm{E}|\eta|^2$$

which holds for any symmetric bilinear forms

$$\phi(\omega, \omega') = \sum_{j,k} c_{jk}\xi_j(\omega)\xi_k(\omega') = \sum_{j,k} c_{jk}\xi_j \otimes \xi_k$$

on $\Omega \times \Omega$ (see (1.8), (1.9)), defining

$$\phi = \phi(\omega, \omega) = \sum_{j,k} c_{jk} \xi_j(\omega) \xi_k(\omega)$$

on Ω and the corresponding

$$\eta = \phi - \mathrm{E}_0 \phi \in H_2(\xi)$$

which represents the Hermite polynomial of degree 2 (with respect to the Gaussian distribution $\mathbf{P}_0$). Then the bijective correspondence

$$\phi(\omega, \omega') \leftrightarrow \eta = \phi - \mathrm{E}_0 \phi \tag{1.13}$$

between these random variables (see (1.7)) can be extended by the limit to all $\eta \in H_2(\xi)$ and the corresponding limits $\phi(\omega, \omega')$ which form a certain space we denote by $H \times H$ with the norm

$$\|\phi\|^2 = \iint |\phi(\omega, \omega')|^2 \, \mathbf{P}(d\omega) \, \mathbf{P}(d\omega').$$

Now, we consider Gaussian distributions $\mathbf{P}$ and $\mathbf{P}_0$ for random variables $\{\xi\}$ labeled by the index $u \in U$ with equal means $A = A_0 = 0$ and different correlation functions B and B_0. We define

$$b(u, v) = B(u, v) - B_0(u, v), \qquad u, v \in U.$$

It turns out that *for the density* $p(\omega) = \mathbf{P}(d\omega)/\mathbf{P}_0(d\omega')$ *the formula*

$$p = \sigma^2 e^{-\frac{1}{2}\eta} \tag{1.14}$$

holds with $\eta = \phi - \mathrm{E}_0 \phi \in H_2(\xi)$ *being uniquely determined by* $\phi(\omega, \omega') \leftrightarrow \eta$ *from the equation*

$$b(u, v) = \iint \xi(\omega, u) \xi(\omega', u) \phi(\omega, \omega') \, \mathbf{P}(d\omega) \, \mathbf{P}_0(d\omega'), \tag{1.15}$$

and the representation (1.15) *with* $\phi(\omega, \omega') \in H \times H$ *is itself a necessary and sufficient condition for the existence of the density* p*; the constant* σ^2 *in* (1.14) *is determined by the normalization*

$$\int p(\omega) \, \mathbf{P}_0(d\omega) = 1.$$

Condition (1.15) is obviously symmetric with respect to $\mathbf{P}$ and $\mathbf{P}_0$: under their rearrangement, only the sign of the difference and of the random variable $\phi(\omega, \omega')$ in (1.15) is changed. This shows that *the Gausssian distributions* **P** *and* $\mathbf{P}_0$ *are either mutually absolutely continuous (equivalent) or mutually singular.*

The easiest way to obtain (1.14) and (1.15) is to turn first to a finite number of orthogonal variables $\{\xi(u),\ u = 1, \dots, n\}$ in $H = H_1(\xi)$ which are orthonormal with respect to the distribution $\mathbf{P}_0$ and orthogonal with respect to **P**,

$$\mathrm{E}\xi^2(u) = \sigma^2(u), \qquad u = 1, \dots, n.$$

On the σ-algebra $\mathcal{A}_n$ generated by $\{\xi(u),\ u = 1, \dots, n\}$, the corresponding density $p_n(\omega) = \mathbf{P}(d\omega)/\mathbf{P}_0(d\omega)$ has the form

$$p_n = \sigma^2 \exp\Big\{-\frac{1}{2}\eta_n\Big\},$$

where $\eta_n = \phi_n - \mathrm{E}\phi_n \leftrightarrow \phi_n(\omega, \omega')$ corresponds, according to (1.13), to the bilinear form

$$\phi_n(\omega, \omega') = \sum_{u=1}^{n} \frac{\sigma^2(u) - 1}{\sigma^2(u)} \xi(\omega, u)\xi(\omega' u).$$

Indeed, for the chosen $\{\xi(u),\ u = 1, \dots, n\}$, we have

$$b(u, v) = \mathrm{E}\xi(u)\xi(v) - \mathrm{E}_0\xi(u)\xi(v) = \begin{cases} \sigma^2 - 1, & v = u, \\ 0, & v \neq u, \end{cases}$$

which coincides with the coefficients of $\phi(\omega, \omega')$ in its expansion over the orthogonal system $\xi(\omega, u)\xi(\omega', v)$, $u, v = 1, \dots, n$, in $H \times H$. For widening sets $\{\xi(u),\ u = 1, \dots, n\}$, the existence of the limit

$$p = \lim p_n$$

is equivalent to existence of the limit $\eta = \lim \eta_n$ and conditions for this limit existence can be expressed in the form (1.15) which immediately leads us to the limit density (1.14).

Actually, the representation (1.15) and the random variable $\eta = \phi - \mathrm{E}_0\phi \in H_2(\xi)$ do not depend on labelling of $\xi = \xi(u)$ by the variable $u \in U$ and (1.15) expresses the fact that the difference in the correlation functions determines a linear continuous functional on the Hilbert space $H \times H$ which on the bilinear form $\phi(\omega, \omega') = \sum_{j,k} c_{jk}\xi(\omega, u_j)\xi(\omega', u_k)$ takes the value $\sum_{j,k} c_{jk} b(u_j, u_k)$. Then,

(1.15) is, in fact, the Riesz representation of this linear functional on the complete system of elements $\xi(\omega, u)\xi(\omega', v)$, $u, v \in U$, in the space $H \times H$. Clearly, representation (1.15) will be valid if, in the above-mentioned sense, the difference in the correlation functions $b(u, v)$ determines a linear continuous functional on the Hilbert space $H \times H$ with an equivalent norm

$$\|\phi\|^2 = \iint |\phi(\omega, \omega')|^2 \, \mathbf{P}_0(d\omega) \, \mathbf{P}_0(d\omega')$$
$$\asymp \iint |\phi(\omega, \omega')|^2 \, \mathbf{P}(d\omega) \, \mathbf{P}_0(d\omega').$$

As already noted, this equivalence for quadratic forms is a consequence of condition (1.12) and by virtue of it (1.15) can be replaced by the analogous representation

$$b(u, v) = \iint \xi(\omega, u)\xi(\omega', v)\phi_0(\omega, \omega') \, \mathbf{P}_0(d\omega) \, \mathbf{P}_0(d\omega') \qquad (1.16)$$

with the corresponding $\phi_0(\omega, \omega') \in H \times H$. Let us emphasize that this representation is equivalent to (1.15) only under the equivalence condition (1.12).

Turning to (1.16), we consider a linear bounded operator $\mathbf{b}$ in the Hilbert space $H = H_1(\xi)$ determined by the bilinear form of $\xi, \eta \in H$

$$\mathrm{E}\xi\eta - \mathrm{E}_0\xi\eta = \langle \xi, \mathbf{b}\eta \rangle_H. \qquad (1.17)$$

From

$$\langle \xi(u), \mathbf{b}\eta(v) \rangle_H = \int_\Omega \xi(\omega, u)[\mathbf{b}\xi(\omega, v)] \, \mathbf{P}_0(d\omega) = \mathbf{b}(u, v)$$
$$= \int_\Omega \xi(\omega, u) \left[\int_\Omega \phi_0(\omega, \omega')\xi(\omega', v) \, \mathbf{P}_0(d\omega') \right] \mathbf{P}_0(d\omega)$$

we see that the operator $\mathbf{b}$ is given by the kernel $\phi_0(\omega, \omega') \in H \times H$ as

$$\mathbf{b}\eta(\omega) = \int_\Omega \phi_0(\omega, \omega')\eta(\omega') \, \mathbf{P}_0(d\omega)$$

with

$$\iint |\phi_0(\omega, \omega')|^2 \, \mathbf{P}_0(d\omega) \, \mathbf{P}_0(d\omega') \le \infty.$$

As a result, we find that *the necessary and sufficient condition for absolute continuity of* $\mathbf{P}$ *with respect to* $\mathbf{P}_0$ *is that the equivalence condition* (1.12) *holds and that the operator* $\mathbf{b}$ *in* (1.17) *determined by the difference of correlations is a Hilbert–Schmidt operator.*

IV.2. Identification of Stochastic Differential Equations

2.1. Equivalence conditions and singularity of Gaussian distributions

Let us turn to the model of a generalized random field $\xi = (x, \xi)$, $x \in D = C_0^\infty(T)$, which was introduced in Section 1.1, Chapter 2. This model represents the unique solution $\xi \in \mathbf{W}$ of the stochastic differential equation

$$\mathcal{L}\xi = \eta$$

in the region $T \subseteq R^d$, where, as we recall, $\mathcal{L}$ can be a differential operator of arbitrary type considered in the space $F = L_2(T)$ or can be a differential operator of type $\mathcal{L} = \mathcal{L}^* = \mathcal{P} \geq 0$ in the corresponding space $F = W = [D]$ with the scalar product

$$\langle u, v \rangle_W = (u, \mathcal{P}v), \qquad u, v \in D;$$

the stochastic sourse will be regarded as being represented by Gaussian white noise on the corresponding F. For different operators $\mathcal{L}$, question is on the equivalence of the corresponding probability distributions for $\xi = (x, \xi)$, $x \in D$.

Here, the question concerns Gausssian distributions with zero mean and the correlation operator B which, as we know (see (1.7), Chapter 3) has the form

$$B = \mathcal{P}^{-1} \tag{2.1}$$

where $\mathcal{P} = \mathcal{L}^*\mathcal{L}$ in the scheme with the operator $\mathcal{L}$ in $F = L_2(T)$ and $\mathcal{P} = \mathcal{L}$ in the scheme with $\mathcal{L} = \mathcal{L}^* = \mathcal{P} \geq 0$ in $F = W = [D]$. Here we recall that $\xi = (x, \xi)$, $x \in D$, represents the unique solution $\xi \in \mathbf{W}$ of the stochastic differential equation

$$\mathcal{P}\xi = \eta \tag{2.2}$$

with the white noise source η on $W = [D]$, and employing of the equation of the general form (2.2) immediately yields the connection with the corresponding probability distribution, indicated in (2.1),

$$B = \mathcal{P}^{-1}\colon X \to W,$$

where $W = X^*$ plays a role of the dual space to the known test functions space $X = [D]$ for $\xi = (x, \xi)$, $x \in X$.

We consider different operators $\mathcal{P}$ and $\mathcal{P}_0$ (associated with W, X and W_0, X_0) denoting by $\mathbf{P}$ and $\mathbf{P}_0$ the Gaussian distributions of $\xi = (x, \xi)$, $x \in D$. The condition of equivalence of $\mathbf{P}$ and $\mathbf{P}_0$ includes the relation

$$\|x\|^2 = \mathrm{E}|(x,\xi)|^2 \asymp \mathrm{E}_0|(x,\xi)|^2 = \|x\|^2_{X_0}, \qquad x \in D, \tag{2.3}$$

and we assume that it holds. Then we have $W = W_0$, $X = X_0$ and with the corresponding operator B,

$$B = \mathcal{P}^{-1}\colon\ X_0 \to W_0.$$

The following necessary and sufficient condition for equivalence of the Gaussian distributions $\mathbf{P}$ and $\mathbf{P}_0$ holds: *the difference*

$$B - B_0\colon\ X_0 \to W_0$$

is a Hilbert–Schmidt operator. And this is equivalent to that

$$(B - B_0)\mathcal{P}_0\colon\ W_0 \to W_0$$

is a Hilbert–Schmidt operator, since

$$\mathcal{P}_0 = B_0^{-1}\colon\ W_0 \to X_0 = \mathcal{P}_0 W_0$$

is the unitary operator. Multiplying the equation

$$(B - B_0)\mathcal{P}_0 = B\mathcal{P}_0 - I$$

from the left by the operator $\mathcal{P}$ (bounded with bounded inverse $B = \mathcal{P}^{-1}$), we obtain the following result.

THEOREM. *The necessary and sufficient condition for equivalence of* $\mathbf{P}$ *and* $\mathbf{P}_0$ *is that the difference*

$$\mathcal{P} - \mathcal{P}_0\colon\ W_0 \to X_0 \tag{2.4}$$

is a Hilbert–Schmidt operator.

Let us consider equation (2.2) with the differential operator

$$\mathcal{P} = \sum_{|k| \le 2p} a_k \partial^k \ge 0 \tag{2.5}$$

on the Sobolev space $W = \overset{\circ}{W}{}^p_2(T)$ with

$$\|u\|^2_W = (u, \mathcal{P}u) \asymp \sum_{|k| \le p} \int \|\partial^k u\|^2_{L_2} = \|u\|^2_p. \tag{2.6}$$

For the operators $\mathcal{P}$ and $\mathcal{P}_0$, their difference looks like

$$\delta\mathcal{P} = \mathcal{P} - \mathcal{P}_0 = \sum_{|k| \le 2p} \delta a_k \partial^k.$$

Let us define $q \le 2p$ such that the coefficients difference $\delta a_k = 0$ for

$$q < |k| \le 2p.$$

The differential operator $\delta\mathcal{P}$ results from the sequence of ∂-differentiations and w-multiplications by $w = \delta a_k$. Obviously, any differentiation

$$\partial\colon \overset{\circ}{W}{}_2^m(T) \to \overset{\circ}{W}{}_2^{m-1}(T)$$

in the Sobolev spaces with $m > 0$ represents the bounded operator and this remains true for $m \le 0$. Indeed, for $m \le 0$

$$\|\partial x\|_{m-1} = \sup_{\|\varphi\|_{-m+1} \le 1} |(\varphi, \partial x)| \le \|x\|_m, \qquad x \in \overset{\circ}{W}{}_2^m(T),$$

because

$$\begin{aligned} |(\varphi, \partial x)| &= |(\partial\varphi, x)| \le \|\partial\varphi\|_{-m}\|x\|_m \\ &\le \|\varphi\|_{-m+1}\|x\|_m, \qquad \varphi \in C_0^\infty(T). \end{aligned}$$

Any w-multiplication

$$w\colon \overset{\circ}{W}{}_2^m(T) \to \overset{\circ}{W}{}_2^m(T)$$

is a bounded operator if the derivatives $|\partial^k w| \le C$, $|k| \le |m|$, are bounded. This is obvious for $m \ge 0$, but holds for $m < 0$ as well. Indeed, for $m < 0$

$$\|wx\|_m = \sup_{\|\phi\|_{-m} \le 1} |(\phi, wx)| \le C\|x\|_m$$

because

$$\begin{aligned} |(\varphi, wx)| &= |(w\varphi, x)| \le \|w\varphi\|_{-m}\|x\|_m \\ &\le C\|\varphi\|_{-m}\|x\|_m, \qquad \varphi \in C_0^\infty(T). \end{aligned}$$

Hence, $\delta\mathcal{P} = \mathcal{P} - \mathcal{P}_0$ is bounded as the operator

$$\delta\mathcal{P}\colon \overset{\circ}{W}{}_2^p(T) \to \overset{\circ}{W}{}_2^m(T), \qquad m = p - q \ge -p.$$

Let us recall that the embedding

$$I\colon \overset{\circ}{W}{}_2^m(T) \to \overset{\circ}{W}{}_2^{-p}(T)$$

is a Hilbert–Schmidt operator in the bounded region $T \subseteq R^d$ if

$$m - (-p) = m + p > \frac{d}{2}$$

and therefore, the scheme

$$\overset{\circ}{W}{}_2^p(T) \xrightarrow{\delta\mathcal{P}} \overset{\circ}{W}{}_2^m(T) \xrightarrow{I} \overset{\circ}{W}{}_2^{-p}(T)$$

with

$$m + p = 2p - q > \frac{d}{2}$$

gives $\delta\mathcal{P} = I \cdot \delta\mathcal{P}$ as the Hilbert–Schmidt operator

$$\delta\mathcal{P}\colon\ \overset{\circ}{W}{}_2^p(T) \to \overset{\circ}{W}{}_2^{-p}(T).$$

Thus, for stochastic differential equations (2.2) with elliptic operators of type (2.5), (2.6) in the bounded region $T \subseteq R^d$, we have the following result.

THEOREM. *The Gaussian probability distributions* $\mathbf{P}$ *and* $\mathbf{P}_0$ *are equivalent if the differential operators* $\mathcal{P}$ *and* $\mathcal{P}_0$ *have their coefficients difference* $\delta a_k = 0$, $|k| > q$, *where*

$$q < 2p - \frac{d}{2}. \tag{2.7}$$

As was mentioned already in the context of Hilbert–Schmidt operators, condition (2.7) cannot be improved. Moreover, as it will be shown below in the probability model (2.2) with the differential operator $\mathcal{P}$ of type (2.5), (2.6), all coefficients a_k, $|k| > q = 2p - d/2$, can be identified with probability 1 by a realization of the Gaussian random field ξ. Hence, for $\mathcal{P}$ and $\mathcal{P}_0$ with non-zero difference of their any kth coeffitients, $|k| > q$, the corresponding Gaussian probability distributions are singular with respect to each other.

As an example of the results related to what was said above, one can consider the Lévy theorem according to which, for the Brownian motion $\xi = \xi(t)$, $t \geq 0$, with the diffusion coefficient σ^2, for almost all realizations (say, on the interval $I = (0, 1)$), we have

$$\sigma^2 = \lim_{n \to \infty} \sum_{m=0}^{n-1} \left| \xi\left(\frac{k+1}{n}\right) - \xi\left(\frac{k}{n}\right) \right|^2$$

(recall that the Brownian motion is a generalized solution $\xi \in \mathbf{W}$ of equation (2.2) with the operator

$$\mathcal{P} = -\frac{1}{\sigma^2}\frac{d^2}{dt^2}$$

on the interval $T = (0, \infty)$).

2.2. Identification of coefficients

Let us turn to the problem of identification of coefficients of the differential operator $\mathcal{P}$ in the stochastic equation (2.2), given a realization of its solution $\xi \in W$. Let us concentrate on the case of constant coefficients, indicating the operator

$$\mathcal{P} = \mathcal{P}(\partial) = \sum_{\{\alpha\}} a_\alpha \partial^\alpha \tag{2.8}$$

with $\alpha = (\alpha_1, \dots, \alpha_d)$ from a given set $\{\alpha\}$ labelling the derivatives ∂^α. The corresponding polynomial is

$$\mathcal{P} = \mathcal{P}(z) = \sum_{\{\alpha\}} a_\alpha z^\alpha \geq 0$$

with $z = i\lambda$ ($i = \sqrt{-1}$), $\lambda \in R^d$.

We turn to the stationary solution $\xi \in \mathbf{W}$ in $T = R^d$, whose existence condition is expressed by the local integrability of the function

$$f(\lambda) = \frac{1}{\mathcal{P}(i\lambda)}, \qquad \lambda \in R^d, \tag{2.9}$$

which represents the spectral density of the generalized stationary Gaussian field $\xi = (x, \xi)$, $x \in C_0^\infty(R^d)$. Note that then we can relax the condition of ellipticity

$$\left| \sum_{|\alpha|=2p} a_\alpha z^\alpha \right|^2 \geq c|z|^{2p}$$

assuming in (2.8) that $a_\beta \neq 0$ for $\beta \in \{\alpha\}$ being the extreme point of the convex hull of all α of order $|\alpha| = 2p$ in R^d and that the polynomial $\mathcal{P}(z)$ is a majorant of its terms:

$$|z^\alpha| \leq \mathcal{P}(z) \tag{2.10}$$

for the set $\{\alpha\}$ we deal with in (2.8).

Before presenting a more precise statement, we can say briefly that each coefficient a_α indexed with $\alpha = (\alpha_1, \dots, \alpha_d)$ in (2.8) of order

$$|\alpha|(\alpha_1 + \dots + \alpha_d) > 2p - \frac{d}{2} \tag{2.11}$$

can be determined by the realization of $\xi = (\varphi, \xi)$, $\varphi \in C_0^\infty(I)$, in any, even very small region $I \subseteq R^d$ (for convenience, we take a unit cube $I = (0,1)^d$) with the help of quadratic functionals of the form

$$S_n = \sum_m \frac{1}{B_n}((Q\varphi_{mn}, \xi)^2 - A_n), \qquad n \to \infty. \tag{2.12}$$

Here, for each multi-index $\alpha = (\alpha_1, \dots, \alpha_d)$ satisfying (2.11) we take a proper differential operator Q, normalizing constants A_n, B_n, and test functions $\varphi_{nm}(t) = \varphi(n^\gamma t - m)$, obtained from $\varphi \in C_0^\infty(I)$ by the indicated transformation with $n^\gamma t = (n^{\gamma_1} t_1, \dots, n^{\gamma_d} t_d)$ for the proper $\gamma = (\gamma_1, \dots, \gamma_d)$ and $m = (m_1, \dots, m_d)$ on the integer-valued lattice with $m_k = 0, \dots, [n^{\gamma_k}] - 1$, $k = 1, \dots, d$.

Let us move on to a more detailed description of the procedure determining the coefficients a_α of the order indicated in (2.11). We start with extreme $\alpha = \beta$, $|\beta| = 2p$, for which we choose a suitable γ such that

$$(\alpha, \gamma) < (\beta, \gamma) \tag{2.13}$$

for all other $\alpha \neq \beta$. Here, for example, we can take γ with components

$$\gamma_k = 1 + \varepsilon\beta_k^0 + o(\varepsilon) \sim 1 + \varepsilon\beta_k^0, \qquad k = 1, \dots, d,$$

where ε is sufficiently small and $\beta^0 = (\beta_1^0, \dots, \beta_d^0)$ determines the hyperplane $\{\lambda \colon (\lambda, \beta^0) = (\beta, \beta^0)\}$ in R^d separating all $\alpha \neq \beta$ with $(\alpha, \beta^0) < (\beta, \beta^0)$,

$$(\alpha, \gamma) \sim |\alpha| + \varepsilon(\alpha, \beta^0) < 2p + \varepsilon(\beta, \beta^0) \sim (\beta, \gamma).$$

For this extreme $\alpha = \beta$, we take in (2.12)

$$Q = \partial^\beta, \qquad A_n = 0, \qquad B_n = n^{(\beta,\gamma)}(\varphi, \partial^\beta\varphi)$$

with the appropriate $\varphi \in C_0^\infty(I)$, $(\varphi, \partial^\beta\varphi) \neq 0$.

In general, using any differential operator with constant coefficients $Q = Q(\partial)$ and the corresponding polynomial $Q = Q(i\lambda)$, for the random variables $(Q\varphi_{mn},\xi)$ in (2.12), having the stationary field ξ with the spectral density (2.9), we obtain

$$\begin{aligned}\mathrm{E}(Q\varphi_{mn},\xi)^2 &= \mathrm{E}(Q\varphi_{0n},\xi)^2\\ &= \int|\tilde{\varphi}_{0n}(\lambda)|^2\frac{|Q(i\lambda)|^2}{\mathcal{P}(i\lambda)}\,d\lambda,\end{aligned}$$

where the Fourier transform $\tilde{\varphi}_{0n}$ is

$$\tilde{\varphi}_{0n}(\lambda) = \frac{1}{(2\pi)^{d/2}}\int e^{i\lambda t}\varphi(n^{\gamma}t)\,dt = n^{-|\gamma|}\tilde{\varphi}(n^{-\gamma}\lambda), \qquad \lambda\in R^d,$$

which can be found with employing the transformation

$$n^{\gamma_k}t_k\to t_k, \qquad k=1,\dots,d,$$

with the Jacobian $n^{|\gamma|}$, $|\gamma|=\gamma_1+\cdots+\gamma_n$. After a similar change of variables

$$n^{-\gamma_k}\lambda_k\to\lambda_k, \qquad k=1,\dots,d,$$

with the Jacobian $n^{-|\gamma|}$, we arrive at the expression

$$\mathrm{E}(Q\varphi_{mn},\xi)^2 = n^{-|\gamma|}\int|\tilde{\varphi}(\lambda)|^2\frac{|Q(i\lambda n^{\gamma})|^2}{\mathcal{P}(i\lambda n^{\gamma})}\,d\lambda.$$

Using

$$\mathcal{P}(i\lambda n^{\gamma}) = \sum_{\{\alpha\}}a_\alpha(i\lambda)^\alpha n^{(\alpha,\gamma)}$$

and $Q(i\lambda n^{\gamma}) = (i\lambda)^\beta n^{(\beta,\gamma)}$, according to inequality (2.13), for $n\to\infty$ we obtain

$$\frac{|Q(i\lambda n^{\gamma})|^2}{\mathcal{P}(i\lambda n^{\gamma})}\sim\frac{1}{a_\beta}(-i\lambda)^\beta n^{(\beta,\gamma)},$$

$$\mathrm{E}(Q\varphi_{mn},\xi)^2\sim n^{-|\gamma|}\frac{1}{a_\beta}\int|\tilde{\varphi}(\lambda)|^2(-i\lambda)^\beta n^{(\beta,\gamma)}\,d\lambda = n^{-|\gamma|}\frac{B_n}{a_\beta}.$$

For S_n in (2.12), where the number of different $m=(m_1,\dots,m_d)$ is $n^{|\gamma|}1+o(1)$, it yields

$$\lim_{n\to\infty}\mathrm{E}S_n = \theta \tag{2.14}$$

with $\theta = 1/a_\beta$.

Let us show that

$$\lim_{n\to\infty}\mathrm{E}(S_n-\theta)^2 = 0. \tag{2.15}$$

We have S_n in (2.12) as the quadratic form of the Gaussian variables labelled as $\xi(u) = (Qu, \xi)$, $u \in D$. In the correspondence

$$(Qu, \xi) = \xi(u) \leftrightarrow \xi^0(u) = (\mathcal{P}u, \xi)$$

for all $u \in D$, thanks to condition (2.10) giving us

$$|Q(z)|^2 \leq C|\mathcal{P}(z)|^2,$$

we have the inequality

$$\mathrm{E}(Qu, \xi)^2 = \int |\tilde{v}(\lambda)|^2 \frac{|Q(i\lambda)|^2}{\mathcal{P}(i\lambda)} \, d\lambda$$

$$\leq C \int |\tilde{u}(\lambda)|^2 \frac{|\mathcal{P}(i\lambda)|^2}{\mathcal{P}(i\lambda)} \, d\lambda = C\mathrm{E}(\mathcal{P}u, \xi)^2.$$

Hence, according to the general comparison theorem on the quadratic formes (see (1.9)),

$$\mathrm{E}(S_n - \mathrm{E}S_n)^2 \leq C\mathrm{E}(S_n^0 - \mathrm{E}S_n^0)^2 \tag{2.16}$$

where

$$S_n^0 = \frac{1}{B_n} \sum_m (\mathcal{P}\varphi_{mn}, \xi)^2$$

Now, having the limit (2.14), it would be enough to prove that

$$\lim_{n \to \infty} \mathrm{E}(S_n^0 - \mathrm{E}S_n^0)^2 = 0.$$

We have

$$S_n^0 - \mathrm{E}S_n^0 = \frac{1}{B_n} \sum_m \eta_{mn}^0$$

with independent random variables

$$\eta_{mn}^0 = (\mathcal{P}\varphi_{mn}, \xi)^2 - \mathrm{E}(\mathcal{P}\varphi_{mn}, \xi)^2$$

since the test functions φ_{mn} have disjoint supports and $\eta = \mathcal{P}\xi$, in our probability model (2.2), is a generalized random field with independent values, more precisely, a generalized Gaussian field with zero mean and the correlation operator $\mathcal{P} = \mathcal{P}(\partial)$.

Taking into account independence and stationarity, after summing over all m of a total number $n^{|\gamma|}(1 + o(1))$ we obtain

$$\mathrm{E}\left[\sum_m \eta_{mn}^0\right]^2 = \sum_m \mathrm{E}[\eta_{mn}^0]^2$$

$$\sim n^{|\gamma|}\mathrm{E}[(\mathcal{P}\varphi_{0n},\xi)^2 - \mathrm{E}(\mathcal{P}\varphi_{0n},\xi)^2]^2$$
$$= n^{|\gamma|}2[\mathrm{E}(\mathcal{P}\varphi_{0n},\xi)^2]^2 = n^{|\gamma|}2\Big[n^{-|\gamma|}\int|\tilde{\varphi}(\lambda)|^2\frac{|\mathcal{P}(i\lambda n^{\gamma})|^2}{\mathcal{P}(i\lambda n^{\gamma})}\,d\lambda\Big]^2$$
$$\le Cn^{-|\gamma|+2\max(\alpha,\gamma)}.$$

Thus, with $1/B_n = O[n^{-(\beta,\gamma)}]$, we obtain

$$\mathrm{E}|S_n - \mathrm{E}S_n|^2 \le Cn^{-2(\beta,\gamma)-|\gamma|+2\max(\alpha,\gamma)} \tag{2.17}$$

where for our $\gamma = (\gamma_1,\ldots,\gamma_d)$ with components $\gamma_k = 1 + O(\varepsilon)$ for small ε we have

$$(\beta,\gamma) = |\beta| + O(\varepsilon), \qquad |\gamma| = d + O(\varepsilon), \qquad \max(\alpha,\gamma) = 2p + O(\varepsilon).$$

We see that if

$$|\beta| > 2p - \frac{d}{2}$$

then, for sufficiently small ε choosen,

$$\mathrm{E}|S_n - \mathrm{E}S_n|^2 \le Cn^{-\delta}, \qquad n \to \infty,$$

with some $\delta > 0$. This result, together with (2.14), proves the limit relation (2.15) and, moreover, we have

$$\lim S_n = \theta \tag{2.18}$$

with probability 1 for $n = 2^r$ with integer $r \to \infty$, say.

By this procedure, we determine all coefficients in (2.8) which correspond to the extreme points $\alpha = \beta^j$, $j = 1, 2, \ldots$, of the convex hull of $\{\alpha\}$, represented in (2.8). Let us consider other coefficients with the index set $\{\alpha\}$ arranged in a proper order (indicated as $\alpha < \beta$, say). Namely, this order requires that for the given β and the corresponding γ with components

$$\gamma_k = 1 + \varepsilon\beta_k + o(\varepsilon), \qquad k = 1, \ldots d,$$

for all $\alpha < \beta$ we have

$$(\alpha,\gamma) = |\alpha| + \varepsilon(\alpha,\beta) + o(\varepsilon) < |\beta| + \varepsilon(\beta,\beta) + o(\varepsilon) = (\beta,\gamma). \tag{2.19}$$

Obviously, keeping this order, we can employ every γ such that (β^j,γ) are different for all extreme points β^j. For the rest of $\{\alpha\}$, $\alpha = \sum_j c_j\beta^j$ with $0 \le c_j < 1$, we have the strict inequality

$$(\alpha,\gamma) = \sum_j c_j(\beta^j,\gamma) < \max_{\{\beta^j\}}(\beta^j,\gamma) = (\beta^*,\gamma).$$

Taking the already determined coefficients a_α, $\alpha > \beta$, including all extreme $\alpha = \beta^j$, we can determine the coefficient a_α for the next $\alpha = \beta$ by employing in (2.12) the differential operator

$$Q = Q(\partial) = \sum_{\alpha > \beta} a_\alpha \partial^\alpha$$

and

$$A_n = \frac{1}{n^{|\gamma|}}\Big(\varphi, \sum_{\alpha>\beta} a_\alpha n^{(\alpha,\gamma)} \partial^\alpha \varphi\Big) = \frac{1}{n^{|\gamma|}} \int |\tilde\varphi(\lambda)|^2 Q(i\lambda n^\gamma)\, d\lambda,$$

$$B_n = n^{(\beta,\gamma)}(\varphi, \partial^\beta \varphi) = n^{(\beta,\gamma)} \int |\tilde\varphi(\lambda)|^2 (-i\lambda)^\beta\, d\lambda.$$

We do not repeat all calculations, which were presented above, and noting that $\mathcal{P}(-z) = \mathcal{P}(z)$ we directly write down

$$\mathrm{E}S_n \sim \frac{1}{B_n} \int |\tilde\varphi(\lambda)|^2 \Big[\frac{|Q(i\lambda n^\gamma)|^2}{\mathcal{P}(-i\lambda n^\gamma)} - Q(i\lambda n^\gamma)\Big]\, d\lambda$$

$$= \frac{1}{B_n} \int |\tilde\varphi(\lambda)|^2 \Big[a_\beta(-i\lambda)^\beta n^{(\beta,\gamma)}$$

$$+ \sum_{\alpha<\beta} a_\alpha(-i\lambda)^\alpha n^{(\alpha,\gamma)}\Big] \frac{\sum_{\alpha>\beta} a_\alpha (i\lambda)^\alpha n^{(\alpha,\gamma)}}{\sum_\alpha a_\alpha (i\lambda)^\alpha n^{(\alpha,\gamma)}}\, d\lambda$$

where the ratio under the integral with the unique most extreme $(\alpha,\gamma) = (\beta^*,\gamma)$, is assymptotically

$$\frac{a_{\beta^*}(i\lambda)^{\beta^*} n^{(\beta^*,\gamma)} + \dots}{a_{\beta^*}(i\lambda)^{\beta^*} n^{(\beta^*,\gamma)} + \dots} \sim 1$$

as $n \to \infty$. Hence, with chosen B_n, according to inequality (2.19) we have

$$\mathrm{E}S_n \sim \alpha_\beta \frac{1}{B_n} n^{(\beta,\gamma)} \int |\tilde\varphi(\lambda)|^2 (-i\lambda)^\beta\, d\lambda = a_\beta.$$

Thus, we obtain the limit relation (2.14) with $\theta = a_\beta$ and, along with it, the relation (2.15) as well, since, in fact, the estimations of variances of the random variables S_n, on which constants A_n have no influence, gives us (2.17). As a consequence, we have the limit relation (2.18) which determines $\theta = a_\beta$.

Let us state our result as follows.

THEOREM. *With probability* 1 *the limit relation* (2.18) *holds with* $\theta = 1/a_\alpha$ *for the extreme* α *and* $\theta = a_\alpha$ *for other* α, $|\alpha| > 2p - d/2$.

In particular, one can see that for large dimensions

$$d > 4p \tag{2.20}$$

the limit relation (2.18) *determines all coefficients* a_α, i.e., determines completely the differential operator

$$\mathcal{P} = \sum_{|\alpha| \le 2p} a_\alpha \partial^\alpha$$

in equation (2.2) and the spectral density

$$f(\lambda) = \frac{1}{\sum_{|\alpha| \le 2p} a_\alpha (i\lambda)^\alpha}$$

of the stationary Gaussian field ξ described by this equation in $T \subseteq R^d$.

We note that in the case of the hypoelliptic operator $\mathcal{P}$ any solution of equation (2.2) can differ from the stationary ξ only by an additional term which, representing a solution of the homogeneous equation, is smooth and does not change the asymptotic behaviour of the corresponding S_n, so we have the same relation (2.18) as in the stationary case.

2.3. On the likelyhood ratio

Turning to the general stochastic model

$$\mathcal{P}\xi = \eta$$

with the various differential operator $\mathcal{P}: W \to X$ and the Gaussian white noise η on the space $W = [D]$ corresponding to $\mathcal{P} \ge 0$, $D \in C_0^\infty(T)$, in the region $T \subseteq R^d$ (see (2.2)), we can give some reasonable formula for the likelyhood ratio (and conditions of its existence) with respect to some $\mathcal{P} = \mathcal{P}_0: W_0 \to X_0$.

Making use of the operator $\mathcal{P}: W \to X$, for any $\mathcal{P}$ and the corresponding W, we introduce the differential operator $\mathcal{P} \times \mathcal{P}_0$ acting on the functions $w(s,t)$ in the region

$$(s,t) \in T \times T \subseteq R^{2d}$$

such that $\mathcal{P}$ and $\mathcal{P}_0$ act on their own variables s and t, respectively. Using the tensor product $W \otimes W_0$ with the dense subset $C_0^\infty(T \times T)$, we have

$$\mathcal{P} \times \mathcal{P}_0\colon W \times W_0 \to X \times X_0$$

as a unitary operator to the dual space $X \times X_0 = (W \times W_0)^*$, according to our general scheme (in Section 3.1, Chapter 1). More precisely, for the elements

$$u \times v = u(s)v(t) \in W \times W_0$$

with components $u, v \in C_0^\infty(T)$, we have

$$(\mathcal{P} \times \mathcal{P}_0)(u \times v) = \mathcal{P}u \times \mathcal{P}_0 v \in X \times X_0$$

with $x = \mathcal{P}u \in X$ and $y = \mathcal{P}_0 v \in X_0$.

Note that, in fact, we have already used the proper tensor space in (1.13), (1.15) considering the density $p(\omega) = \mathbf{P}(d\omega)/\mathbf{P}_0(d\omega)$ which involves random variables $\phi(\omega, \omega')$ on the product $\Omega \times \Omega$; with obvious changes of notations, in applications to the set of Gaussian variables

$$(x, \xi), \qquad x \in D = C_0^\infty(T),$$

the indicated $\phi(\omega, \omega')$ are

$$\phi(\omega, \omega') = \sum_{j,k} c_{jk}(x_j, \xi_\omega)(x_k, \xi_{\omega'})$$

and their limits in the space $H \times H$ related to (1.13). For equation (2.2) with the stochastic source η being the Gaussian white noise on the corresponding $W = [D]$ when we have

$$\|(x, \xi)\|^2 = \mathrm{E}|(x, \xi)|^2 = \|x\|_X^2 = \|\mathcal{P}u\|_W^2, \qquad x = \mathcal{P}u, \quad u \in W,$$

for all $x \in X = \mathcal{P}W$, we get the isometry

$$\phi(\omega, \omega') \leftrightarrow \sum_{j,k} c_{jk}(x_j \times x_k) \in X \times X_0. \tag{2.21}$$

We introduce

$$(z, \xi_\omega \times \xi_{\omega'}) = \phi(\omega, \omega'), \qquad z = \sum_{j,k} c_{jk}(x_j \times x_k).$$

to employ these elements according to the general scheme (1.13),

$$(z, \xi_\omega \times \xi_{\omega'}) \leftrightarrow \eta = \phi - \mathrm{E}_0\phi \in H_2(\xi). \tag{2.22}$$

Below, for random variables $\eta \in H_2(\xi)$ this isomorphism will be indicated as

$$\eta = (z, \xi \times \xi), \qquad z \in X \times X_0.$$

According to (2.22), the representation (1.15) for the difference in the correlation operators $\mathbf{b} = B - B_0$ gives us

$$(x, \mathbf{b}y) = \iint (x \times y, \xi_\omega \times \xi_{\omega'})(z, \xi_\omega \times \xi_{\omega'})\, \mathbf{P}(d\omega) \times \mathbf{P}_0(d\omega')$$

$$= \langle x \times y, z \rangle_{X \times X_0}, \qquad x, y \in D = C_0^\infty(T),$$

with the complete system $\{x \times y\}$ in $X \times X_0$. This can be interpreted as that the difference $\mathbf{b} = B - B_0$ determines the generalized function $\mathbf{b} \in W \times W_0$ defined as

$$(x \times y, \mathbf{b}) \equiv (x, \mathbf{b}y) = \langle x \times y, z \rangle_{X \times X_0},$$

$$x \times y = x(s)y(t) \in C_0^\infty(T \times T).$$

Clearly, with this interpretation we have

$$z = (\mathcal{P} \times \mathcal{P}_0)\mathbf{b}.$$

Thus, in the framework of this scheme *the criterion for equivalence of Gaussian* $\mathbf{P}$ *and* $\mathbf{P}_0$ *can be given in terms of the difference* $\mathbf{b} = B - B_0$ *by the condition*

$$\mathbf{b} = \{(x \times y, \mathbf{b}) \equiv (x, \mathbf{b}y),\ x \times y \in C_0^\infty(T \times T)\} \in W \times W_0 \quad (2.23)$$

and, when this condition holds,

$$p(\omega) = \frac{\mathbf{P}(d\omega)}{\mathbf{P}_0(d\omega)} = \sigma^2 e^{-\frac{1}{2}(z, \xi \times \xi)}, \quad (2.24)$$

$$z = (\mathcal{P} \times \mathcal{P}_0)\mathbf{b}.$$

IV.3. Estimation of Mean Value in Stochastic Differential Equations

3.1. The best unbiased estimators

Let us stay with the generalized random field $\xi = (\varphi, \xi)$, $\varphi \in C_0^\infty(S)$, uniquely described in the region $T \subseteq R^d$ by the stochastic partial differential equation

$$\mathcal{L}\xi = \eta.$$

One can imagine, that the main object of interest is the solution $u \in W$ of the corresponding deterministic differential equation

$$\mathcal{L}u = f \tag{3.1}$$

which, as a consequence of stochastic perturbations, turns out to be hidden in the random field ξ as

$$u = \mathrm{E}\xi.$$

Here, while dealing with observable ξ, we must estimate the unknown mean value $u \in W$.

We consider the mean $u = \theta$ as the unknown functional parameter of the corresponding Gaussian distribution $\mathbf{P}_\theta$ of

$$\xi = (x, \xi), \qquad x \in X(S) = [C_0^\infty(T)],$$

with the known test functions $x \in X$, assuming that $\theta \in \Theta$ belongs to a certain set $\Theta \subseteq W(S)$ in the functional space W such that along with any $\theta_1, \dots, \theta_n$ in Θ the set Θ contains all $\theta = \sum_{k=1}^n \lambda_k \theta_k$ with $(\lambda_1, \dots, \lambda_n)$ in some n-dimensional parallelipiped in R^n. For $\theta = 0$, the Gaussian random variables (x, ξ), $x \in X$, represent stochastic perturbations which, in general, are imposed on the solution $u = (x, u)$, $x \in X$, of equation (3.1). Turning now to its probability distribution $\mathbf{P}_0$ with the zero mean $\mathrm{E}_0(x, \xi) \equiv 0$, we note that all $\mathbf{P}_\theta$ differ from $\mathbf{P}_0$ only by the corresponding mean values

$$(x, \xi) = \mathrm{E}_\theta(x, \xi), \qquad x \in X.$$

We assume that the stochastic perturbations (with respect to P_0 represented by the very random variables (x, ξ), $x \in X$) are similar to the white noise on the test functions space X, being such that

$$\mathrm{E}_0(x, \xi) = 0,$$

$$\mathrm{E}_0|(x, \xi)|^2 \asymp \|x\|_X^2, \qquad x \in X. \tag{3.2}$$

We use the distribution

$$\mathbf{P} = \mathbf{P}_0$$

and the corresponding Hilbert space $\mathbf{H} = \mathbf{H}(\mathcal{A})$ of the random variables measurable with respect to the σ-algebra $\mathcal{A}$ generated by $\xi = (x, \xi)$, $x \in X$, applying expectation $\mathrm{E} = \mathrm{E}_0$. By virtue of the equivalence condition (3.2), all distributions $\mathbf{P}_\theta$ have the probability density with respect to $\mathbf{P} = \mathbf{P}_0$

$$p_\theta(\omega) = \frac{\mathbf{P}_\theta(d\omega)}{\mathbf{P}(d\omega)} = \sigma_\theta^2 e^{\eta_\theta} \in \mathbf{H} \tag{3.3}$$

— see (1.10). Given the corresponding random variables $\eta_\theta \in H = H_1(\xi)$, we have

$$(x, \theta) = \mathrm{E}(x, \xi)\eta_\theta, \qquad x \in X, \tag{3.4}$$

cf. the general formula (1.11). Actually, in (3.4) we have the expectation $(x, \theta) = \mathrm{E}_\theta \eta$ of the random variables $\eta = (x, \xi)$, $x \in X$, from $H = H_1(\xi)$ in $\mathbf{H}$, which leads to

$$\mathrm{E}_\theta \eta = \mathrm{E}\eta\eta_\theta, \qquad \eta \in H. \tag{3.4$'$}$$

Clearly, according to (3.3), any $\eta \in \mathbf{H}$ has a finite expectation

$$\mathrm{E}_\theta \eta = \mathrm{E}\eta p_\theta, \qquad \theta \in \Theta.$$

Denoting the σ-algebra generated by all random variables η_θ, $\theta \in \Theta$, by $\mathcal{B}$, we note that the set $\{\eta_\theta\}$ forms a *sufficient statistic*, with respect to which the conditionall expectation

$$\mathrm{E}_\theta(\eta|\mathcal{B}) = \mathrm{E}(\eta|\mathcal{B})$$

of any random variable $\eta \in \mathbf{H}$ does not depend on $\theta \in \Theta$. Indeed

$$\int_B \mathrm{E}(\eta|\mathcal{B})\, \mathbf{P}_\theta(d\omega) = \int_B \mathrm{E}(\eta|\mathcal{B}) p_\theta\, \mathbf{P}(d\omega)$$

$$= \int_B \mathrm{E}(\eta p_\theta|\mathcal{B})\, \mathbf{P}(d\omega) = \int_B \eta p_\theta\, \mathbf{P}(d\omega) = \int_B \eta\, \mathbf{P}_\theta(d\omega)$$

for all $B \in \mathcal{B}$, so that $\mathrm{E}_\theta(\eta|\mathcal{B})$ is equal to the random variable $\hat{\eta} = \mathrm{E}(\eta|\mathcal{B})$.

For arbitrary $x \in X$, as a possible estimator of the unknown value (x, ξ), one can use the random variables $\eta \in \mathbf{H}$ having the property of *unbiasedness*

$$\mathrm{E}_\theta \eta = (x, \theta), \qquad \theta \in \Theta;$$

for example, one can use $\eta = (x, \xi)$.

As is known, for any unbiased estimator $\eta \in \mathbf{H}$ there exists the *improved unbiased estimator*

$$\hat{\eta} = \mathrm{E}(\eta|\mathcal{B}) = \mathrm{E}_\theta(\eta|\mathcal{B})$$

which is better than the original η in the sense that the mean square error in the estimate $\hat{\eta}$ is smaller,

$$\mathrm{E}_\theta|\eta - (x,\theta)|^2 = \mathrm{E}_\theta|\hat{\eta} - (x,\theta)|^2 + \mathrm{E}_\theta|\hat{\eta} - \eta|^2 \geq \mathrm{E}_\theta|\hat{\eta} - (x,\theta)|^2.$$

The improved estimator $\hat{\eta}$ is measurable with respect to the σ-algebra $\mathcal{B}$, more precisely

$$\hat{\eta} = \mathrm{E}(\eta|\mathcal{B}) \in \mathbf{H}(\mathcal{B}) \tag{3.5}$$

belongs to the subspace $\mathbf{H}(\mathcal{B})$ of all $\mathcal{B}$-measurable random variables in $\mathbf{H}$. By virtue of the *condition of the completeness* of the sufficient statistic $\{\eta_\theta\}$, which, in the framework of this scheme, means the completeness of random variables p_θ, $\theta \in \Theta$, in the subspace $\mathbf{H}(\mathcal{B})$, there is a unique unbiased estimator $\eta \in \mathbf{H}(\mathcal{B})$, because for the difference of estimators of this kind $\eta = \eta_1, \eta_2$ we obtain

$$\mathrm{E}_\theta(\eta_1 - \eta_2) = \mathrm{E}(\eta_1 - \eta_2)p_\theta = 0, \qquad \theta \in \Theta,$$

which, by virtue of the completeness of $\{p_\theta\}$, implies $\eta_1 = \eta_2$ in $\mathbf{H}(\mathcal{B})$.

The condition imposed on the set $\Theta \subseteq W$ indicates that in the exponential family (3.3) together with each linearly independent $\eta_\theta = \eta_1, \ldots, \eta_n$ thare are $\eta_\theta = \sum_{k=1}^n \lambda_k \eta_k$ with $(\lambda_1, \ldots, \lambda_n)$ in some n-dimensional parallelipiped in R^n and this yields completeness of the exponential family of random variables $\{p_\theta\}$ of the form (3.3) in $\mathbf{H}(\mathcal{B})$. This follows, for example, from completeness of all $\exp\{i\sum_{k=1}^n \lambda_k\eta_k\}$ with $(\lambda_1, \ldots, \lambda_n) \in R^n$ in the complex space $\mathbf{H}(\mathcal{B})$. Indeed, for any random variable $\eta \in \mathbf{H}(\mathcal{B})$, the expression $\mathrm{E}\eta\exp\{i\sum_{k=1}^n z_k\eta_k\}$ defines an analytic function of complex variables $(z_1, \ldots, z_n)$ which is equal to zero identically (and, in particular, in $z_1 = i\lambda_1, \ldots, z_n = i\lambda_n$) if it is equal to zero in some n-dimensional parallelipiped in R^n.

Thus, among all unbiased estimators $\eta \in \mathbf{H}$, we have a unique estimator $\eta = \hat{\eta}$ in $\mathbf{H}(\mathcal{B})$ for the unknown value of (x,θ) which, as was indicated in (3.5), can be obtained from any other unbiased estimator η by averaging it with respect to $\mathcal{B}$; hence, this $\hat{\eta}$ represents *the best estimator*

$$\hat{\eta} = (x, \hat{\theta})$$

with the smallest mean square error

$$\mathrm{E}|(x, \hat{\theta}) - (x, \theta)|^2 = \min .$$

Starting from the trivial unbiased estimator $\eta = (x, \xi)$ and finding the best estimator

$$(x, \hat{\theta}) = \mathrm{E}[(x, \xi)/\mathcal{B}]$$

by averaging with respect to the σ-algebra $\mathcal{B}$ generated by the Gaussian set $\{\eta_\theta\}$ from (3.3), we conclude that $(x, \hat{\theta})$ belongs to the linear closure $H(\mathcal{B})$ of random variables η_θ, $\theta \in \Theta$, and in this sense the best estimator is *linear*

$$(x, \hat{\theta}) \in H(\mathcal{B}).$$

With general representation (3.4) we obtain the following result.

THEOREM. *The best unbiased estimator for (x, θ) is uniquely determined as $(x, \hat{\theta}) \in H(\mathcal{B})$ such that*

$$\mathrm{E}(x, \hat{\theta})\eta_\theta = (x, \theta), \qquad \theta \in \Theta. \tag{3.6}$$

Note that under the condition imposed on the set $\Theta \subseteq W(S)$, the best unbiased estimator $(x, \hat{\theta})$ for (x, θ), $\theta \in \Theta$, is the best unbiased estimator for (x, θ) also in situation when instead of Θ we consider its linear closure in the space W. It is convenient to pass to a case of the parameter set Θ being a subspace $\Theta \subseteq W$ when any $\eta \in H(\mathcal{B})$ serves as $\eta = \eta_\theta$ for $\theta \in \Theta$

$$(x, \theta) = \mathrm{E}(x, \xi)\eta, \qquad x \in X,$$

(see (3.4)). ■

Let us consider the problem conscerning the best estimator of the unknown mean $u = \theta$ in (3.1) under stochastic perturbations (x, ξ), $x \in X$, of the type of white noise on X, i.e.,

$$\mathrm{E}(x, \xi) = 0, \qquad \mathrm{E}|(x, \xi)|^2 = \sigma^2 \|x\|_X^2, \qquad x \in X. \tag{3.7}$$

Let us recall that the test functions space represents in general scheme

$$X = \mathcal{P}W$$

in terms of the operator $\mathcal{P}$, $\mathcal{P} = \mathcal{L}^*\mathcal{L}$, with the general operator $\mathcal{L}$ in the space $F = L_2$, or $\mathcal{P} = \mathcal{L}$ with positive $\mathcal{L} = \mathcal{L}^* = \mathcal{P} \geq 0$ in the corresponding $F = W$. According to (3.7), we have

$$\mathrm{E}(x,\xi)(\mathcal{P}u,\xi) = \sigma^2\langle x, \mathcal{P}u\rangle_X = \sigma^2(x,u), \qquad x \in X,\ u \in W.$$

One can immediately see that the random variables η_θ in representation (3.4) are

$$\eta_\theta = \frac{1}{\sigma^2}(\mathcal{P}\theta,\xi), \qquad \theta \in \Theta. \tag{3.8}$$

Taking an orthonormal basis $\{\theta_k\}$ in the subspace $\Theta \subseteq W$, with the help of the unitary operator $\mathcal{P}: W \to X$, we obtain the orthonormal system $\{\mathcal{P}\theta_k\}$ in the space X with the corresponding basis $\{\frac{1}{\sigma^2}(\mathcal{P}\theta_k,\xi)\}$ in the subspace $H(\mathcal{B})$ of random variables η_θ, $\theta \in \Theta$. Using the expansion

$$(x,\hat{\theta}) = \sum_k c_k(\mathcal{P}\theta_k,\xi)$$

of the estimator $(x,\hat{\theta})$, from (3.6) we find the coefficients

$$c_k = \mathrm{E}(x,\hat{\theta})\Big[\frac{1}{\sigma^2}(\mathcal{P}\theta_k,\xi)\Big] = (x,\theta_k).$$

Thus, for estimators of the solution $u = \theta \in \Theta$ of equation (3.1) with its values $(x,u) = (x,\theta)$, $x \in X$, under stochastic perturbations of the white noise type on the test functions space X, we obtain the following result.

THEOREM. *The best estimator $(x,\hat{\theta})$ can be given with the expansion*

$$(x,\hat{\theta}) = \sum_k (x,\theta_k)(\mathcal{P}\theta_k,\xi), \qquad x \in X, \tag{3.9}$$

over the orthonormal basis $\{\theta_k\}$ in $\Theta \subseteq W$.

By representing functional series (3.9) with random coefficients $\eta_k = (\mathcal{P}\theta_k,\xi)$ in the form

$$(x,\hat{\theta}) = \sum_k \eta_k(x,\theta_k), \qquad x \in X,$$

for the finite-dimensional parameter space Θ, we obtain the *functional estimator*

$$\hat{\theta} = \sum_k \eta_k\theta_k \in \Theta \tag{3.10}$$

of the unknown solution $u = \theta$ in (3.1). Unfortunately, the functional expansion (3.10) over the orthonormal basis $\{\theta_k\}$ in Θ does not work for infinite-dimensional Θ, since $\sum_k |\eta_k|^2 = \infty$ with probability 1 for independent Gaussian random variables $\eta_k = (\mathcal{P}\theta_k, \xi)$, $\mathrm{E}|\eta_k|^2 = \sigma^2$; here we can only guarantee with probability 1 the convergence of the series $\sum_k \eta_k(x, \theta_k)$ for which at any $x \in X$ we have

$$\mathrm{E}\sum_k |\eta_k(x,\theta_k)|^2 = \sigma^2 \sum_k |(x,\theta_k)|^2 \le \sigma^2 \|x\|_X^2.$$

Let us also note that formula (3.9) for the best estimator does not depend on the factor σ^2 in (3.7) which characterizes the intensity of the stochastic perturbations of the white noise type.

3.2. Pseudo-best estimators, the method of east squares, and the consistency condition

It is easy to imagine the situation when we are forced to deal with the unknown probability distribution $\mathbf{P}$ of stochastic perturbatins, e.g. it is only known that for this distribution the mean $\mathrm{E}(x, \xi) \equiv 0$ and the random variables (x, ξ) are mean square continuous with respect to $x \in X$ in the test functions space X, i.e.,

$$\mathrm{E}|(x,\xi)|^2 \le C\|x\|_X^2, \qquad x \in X.$$

Regardless of what the present probability distribution $\mathbf{P}$ is, we can try to use estimator which is the best with respect to some probability known distribution $\widehat{\mathbf{P}}$. For example, considering a generalized solution $u = \theta \in W$ of equation (3.1) we can employ the embedding

$$W \subseteq \widehat{W} \tag{3.11}$$

regarding $u = (x, u)$ as a function of $x \in \widehat{X}$ in the corresponding test functions space

$$\widehat{X} = \widehat{W}^* \subseteq W^* = X$$

and, in terms of the pseudo-distribution $\widehat{\mathbf{P}}$ of $\xi = (x, \xi)$, $x \in \widehat{X}$, take the white noise on the space $\widehat{X}$.

According to this, as a proper estimator for unknown $u = \theta \in \Theta$ in $\widehat{W}$, we suggest $\hat{\theta}$ defined by the formula (3.9) with the orthonormal basis $\{\theta_k\}$ in $\Theta \subseteq \widehat{W}$ and the operator $\mathcal{P}$ replaced by the corresponding operator

$$\widehat{\mathcal{P}}\colon \widehat{W} \to \widehat{X}$$

which yields the obtained in this way $\hat{\theta}$:

$$(x,\hat{\theta}) = \sum_k (x,\theta_k)(\widehat{\mathcal{P}}\theta_k,\xi); \tag{3.12}$$

we call $\hat{\theta}$ the *pseudo-best estimator* — this estimator is the best in a case of $\widehat{\mathbf{P}} = \mathbf{P}$.

THEOREM. *The pseudo-best estimators are unbiased.*

Indeed,

$$\mathrm{E}_\theta(x,\hat{\theta}) = \sum_k (x,\theta_k)\mathrm{E}_\theta(\widehat{\mathcal{P}}\theta_k,\xi) = \sum_k (x,\theta_k)(\widehat{\mathcal{P}}\theta_k,\theta) = (x,\theta) \tag{3.13}$$

since for $\theta \in \Theta$ in the Hilbert space $\widehat{W}$ we have the expansion

$$\theta = \sum_k (\widehat{\mathcal{P}}\theta_k,\theta)\theta_k$$

with respect to the orthonormal basis $\{\theta_k\}$ in $\Theta \subseteq \widehat{W}$.

In addition, let us consider the case when (3.11) is Hilbert–Schmidt embedding. Then, the generalized random function $\xi = (x,\xi)$, $x \in \widehat{X}$, observable as the result of stochastic perturbations of the solution $u = \theta \in \widehat{W}$ in (3.1), has an equivalent modification all realizations of which are

$$\xi_\omega \in \widehat{W} \tag{3.14}$$

(see Section 2.1, Chapter 1). Regarding the realization $\xi \in \widehat{W}$ as an estimate of the hidden component $u = \theta$, we can try to improve it with the best approximation for $\xi \in \widehat{W}$ by functions in the parameter space $\Theta \ni \theta = u$, more precisely, in its closure $[\Theta]$ in $\widehat{W}$. This best approximation $\hat{\theta} \subseteq [\Theta]$ is given by the projection of the element ξ onto the subspace $[\Theta]$ in the Hilbert space $\widehat{W}$. Taking an orthonormal basis $\{\theta_k\}$ in $\Theta \subseteq \widehat{W}$ and supplementing it by elements $\{u_k\}$ to an orthonormal basisi in the whole $\widehat{W}$, with the expansion

$$\xi = \sum_k (\widehat{\mathcal{P}}\theta_k,\xi)\theta_k + \sum_k (\widehat{\mathcal{P}}u_k,\xi)u_k$$

of the element $\xi \in \widehat{W}$ over this basis we obtain

$$\hat{\theta} = \sum_k (\widehat{\mathcal{P}}\theta_k,\xi)\theta_k$$

which, for all $x \in \widehat{X}$, gives

$$(x, \hat{\theta}) = \sum_k (x, \theta_k)(\widehat{\mathcal{P}}\theta_k, \xi).$$

It is easy to see that the *method of least squares* presented above gives the estimates $(x, \hat{\theta})$ introduced by the formula (3.12) as the pseudo-best ones (the best for the distribution $\widehat{\mathbf{P}} = \mathbf{P}$). We formulate this result in the form of the following proposition.

THEOREM. *The least-squares estimators are pseudo-best.*

Let us return to the best estimators obtained by the suggested method for $\widehat{\mathbf{P}} = \mathbf{P}$. Clearly, even for the best estimators $(x, \hat{\theta})$, the question arises as how good they are from the point of view of approximation to the estimated values (x, θ), e.g. in a sense of the meansquare distance

$$\mathrm{E}_\theta|(x, \hat{\theta}) - (x, \theta)|^2 = \mathrm{E}|(x, \hat{\theta})|^2. \tag{3.15}$$

Suppose, we are getting data concerning the random field $\xi = (x, \xi)$ with $x \in X$ in enlarging regions $T = T_n$, $n = 1, 2, \ldots$, and correspondingly obtain the estimators $\hat{\theta} = \hat{\theta}^{(n)}$ in each region T_n. Then, the estimators $\hat{\theta}$ can be characterized by their *consistency*, which means that

$$\mathrm{E}_\theta|(x, \hat{\theta}) - (x, \theta)|^2 \to 0 \tag{3.16}$$

as $n \to \infty$.

Here, the generalized functions $u = \theta \in W = W_n$ defined in each region $T = T_n$ can be regarded all together as generalized function in the region $T_\infty = \cup_n T_n$. To be more precise, we assume that the parameter space Θ is represented by the generalized functions

$$\theta = (\varphi, \theta), \qquad \varphi \in C_0^\infty(T_\infty),$$

in the region $T_\infty = \cup_n T_n$, the restriction of which in any region $T = T_n$ is $\theta = \theta_n \in W_n$.

In any region $T = T_n$, we have $\theta \in W_n$ with the norm

$$\|\theta\|_{W_n} = \sup_{\|x\|_{X_n}=1} (x, \theta)$$

where sup is taken over the corresponding $x \in X = X_n$ and the boundedness of this norm for θ in all $T = T_n$ means that $\theta \in W$ in the region $T = T_\infty$ which is exactly the known condition of equivalence of

Gaussian distributions $\mathbf{P}_\theta$ and $\mathbf{P} = \mathbf{P}_0$ for (x, ξ), $x \in X$, in $T = T_\infty$. Hence, convergence in (3.16) can, obviously, hold only for $\theta = 0$. Thus, the condition

$$\|\theta\|_{W_n} \to \infty, \qquad n \to \infty, \tag{3.17}$$

for $\theta \neq 0$ is necessary for consistency of the best estimators.

Let us consider the one-dimensional parameter space Θ. Then, according to the general formula (3.9) with only $k = 0$ and

$$\theta_0 = \frac{\theta}{\|\theta\|_{W_n}}$$

we have

$$(x, \hat{\theta}) = \frac{1}{\|\theta\|_{W_n}} (x, \theta)(\widehat{\mathcal{P}}\theta, \xi)$$

and

$$\mathbf{E}_\theta |(x, \hat{\theta}) - (x, \theta)|^2 = \mathbf{E}|(x, \hat{\theta})|^2$$

$$\leq \frac{C}{\|\theta\|^4_{W_n}} |(x, \theta)|^2 \|\widehat{\mathcal{P}}\theta\|^2_{X_n} = C\|\theta\|^2_{W_n} |(x, \theta)|^2 \to 0$$

as $n \to \infty$, if condition (3.17) holds.

Let us show that this condition in general implies consistency of the estimates $(x, \hat{\theta})$, $n \to \infty$.

We assume that the parameter space Θ of the generalized functions $\theta = (\varphi, \theta)$, $\varphi \in T_\infty$, in the region $T_\infty = \cup T_n$, is closed with respect to weak convergence.

Starting with an unbiased estimator in the region $T = T_n$ and using the corresponding projection operation described in its general form in (3.5), we can obtain the best estimator in the enlarged region $T = T_{n+1} \supseteq T_n$. Thus, for any $x \in X$ in $T = T_\infty$ with support in some region T_m, the sequence of best estimators $\eta^{(n)} = (x, \hat{\theta}^{(n)})$, $n \geq m$, can be obtained by succesive projections there is the limit $\eta = \lim \eta^{(n)}$ in $\mathbf{H}$. As we know, the expectation

$$\mathbf{E}(\varphi, \xi)\eta^{(n)} = (\varphi, \theta^{(n)}), \qquad \varphi \in C_0^\infty(S_n),$$

determines a function $\theta^{(n)} \in \Theta$ in the region T_n. Obviously, by virtue of the convergence of $\eta^{(n)}$ in $\mathbf{H}$, we have weak convergence of the generalized functions $\theta^{(n)} \in \Theta$ in the region $T_\infty = \cup T_n$ and from the fact that Θ is closed, we obtain the limit function $\theta = \lim \theta^{(n)} \in \Theta$ for which

$$\mathrm{E}(\varphi, \xi)\eta = (\varphi, \theta), \qquad \varphi \in C_0^\infty(T_\infty).$$

The representation of the function $\theta \in \Theta$ obtained above indicated its continuity with respect to $\varphi \in C_0^\infty(T_\infty)$ in the norm $\|\phi\|_X$ which means that $\theta \in W$ in $T = T_\infty$ and by virtue of condition (3.17), this can only happen if $\theta = 0$ and $\eta = 0$. To conclude, we see from condition (3.17) that

$$\lim \eta^{(n)} = \eta = 0$$

in $\mathbf{H}$, i.e.,

$$\mathrm{E}|\eta^{(n)}|^2 = \mathrm{E}_\theta|\eta^{(n)} - (x, \theta)|^2 \to 0,$$

as $n \to \infty$.

Passing to the pseudo-best estimators $(x, \hat{\theta})$ in the scheme suggested (with $\widehat{W} \supseteq W$, $\widehat{X} \subseteq X$ and the corresponding pseudo-meansquare norm stronger then the true one for $\eta \in \mathbf{H}$ from the closure of all (x, ξ), $x \in X$, in $T = T_\infty$), we obtain the following condition of consistency.

THEOREM. *If for all functions $\theta \neq 0$*

$$\|\theta\|_{\widehat{W}_n} \to \infty$$

then the pseudo-best estimators $(x, \hat{\theta})$ for all values (x, θ) are consistent.

Other *Mathematics and Its Applications* titles of interest:

A.V. Ivanov and N.N. Leonenko: *Statistical Analysis of Random Fields.* 1989, 256 pp. ISBN 90-277-2800-3

V. Paulauskas and A. Rackauskas: *Approximation Theory in the Central Limit Theorem. Exact Results in Banach Spaces.* 1989, 176 pp. ISBN 90-277-2825-9

R.Sh. Liptser and A.N. Shiryayev: *Theory of Martingales.* 1989, 808 pp. ISBN 0-7923-0395-4

S.M. Ermakov, V.V. Nekrutkin and A.S. Sipin: *Random Processes for Classical Equations of Mathematical Physics.* 1989, 304 pp. ISBN 0-7923-0036-X

G. Constantin and I. Istratescu: *Elements of Probabilistic Analysis and Applications.* 1989, 488 pp. ISBN 90-277-2838-0

S. Albeverio, Ph. Blanchard and D. Testard (eds.): *Stochastics, Algebra and Analysis in Classical and Quantum Dynamics.* 1990, 264 pp. ISBN 0-7923-0637-6

Ya.I. Belopolskaya and Yu.L. Dalecky: *Stochastic Equations and Differential Geometry.* 1990, 288 pp. ISBN 90-277-2807-0

A.V. Gheorghe: *Decision Processes in Dynamic Probabilistic Systems.* 1990, 372 pp. ISBN 0-7923-0544-2

V.L. Girko: *Theory of Random Determinants.* 1990, 702 pp. ISBN 0-7923-0233-8

S. Albeverio, PH. Blanchard and L. Streit: *Stochastic Processes and their Applications in Mathematics and Physics.* 1990, 416 pp. ISBN 0-9023-0894-8

B.L. Rozovskii: *Stochastic Evolution Systems. Linear Theory and Applications to Non-linear Filtering.* 1990, 330 pp. ISBN 0-7923-0037-8

A.D. Wentzell: *Limit Theorems on Large Deviations for Markov Stochastic Process.* 1990, 192 pp. ISBN 0-7923-0143-9

K. Sobczyk: *Stochastic Differential Equations. Applications in Physics, Engineering and Mechanics.* 1991, 410 pp. ISBN 0-7923-0339-3

G. Dallaglio, S. Kotz and G. Salinetti: *Distributions with Given Marginals.* 1991, 300 pp. ISBN 0-7923-1156-6

A.V. Skorohod: *Random Processes with Independent Increments.* 1991, 280 pp. ISBN 0-7923-0340-7

L. Saulis and V.A. Statulevicius: *Limit Theorems for Large Deviations.* 1991, 232 pp. ISBN 0-7923-1475-1

A.N. Shiryaev (ed.): *Selected Works of A.N. Kolmogorov, Vol. 2: Probability Theory and Mathematical Statistics.* 1992, 598 pp. ISBN 90-277-2795-X

Yu.I. Neimark and P.S. Landa: *Stochastic and Chaotic Oscillations.* 1992, 502 pp. ISBN 0-7923-1530-8

Other *Mathematics and Its Applications* titles of interest:

Y. Sakamoto: *Categorical Data Analysis by AIC*. 1992, 260 pp. ISBN 0-7923-1429-8

Lin Zhengyan and Lu Zhuarong: *Strong Limit Theorems*. 1992, 200 pp. ISBN 0-7923-1798-0

J. Galambos and I. Katai (eds.): *Probability Theory and Applications*. 1992, 350 pp. ISBN 0-7923-1922-2

N. Bellomo, Z. Brzezniak and L.M. de Socio: *Nonlinear Stochastic Evolution Problems in Applied Sciences*. 1992, 220 pp. ISBN 0-7923-2042-5

A.K. Gupta and T. Varga: *Elliptically Contoured Models in Statistics*. 1993, 328 pp. ISBN 0-7923-2115-4

B.E. Brodsky and B.S. Darkhovsky: *Nonparametric Methods in Change-Point Problems*. 1993, 210 pp. ISBN 0-7923-2122-7

V.G. Voinov and M.S. Nikulin: *Unbiased Estimators and Their Applications. Volume 1: Univariate Case*. 1993, 522 pp. ISBN 0-7923-2382-3

V.S. Koroljuk and Yu.V. Borovskich: *Theory of* U-*Statistics*. 1993, 552 pp. ISBN 0-7923-2608-3

A.P. Godbole and S.G. Papastavridis (eds.): *Runs and Patterns in Probability: Selected Papers*. 1994, 358 pp. ISBN 0-7923-2834-5

Yu. Kutoyants: *Identification of Dynamical Systems with Small Noise*. 1994, 298 pp. ISBN 0-7923-3053-6

M.A. Lifshits: *Gaussian Random Functions*. 1995, 346 pp. ISBN 0-7923-3385-3

M.M. Rao: *Stochastic Processes: General Theory*. 1995, 635 pp. ISBN 0-7923-3725-5

Yu.A. Rozanov: *Probability Theory, Random Processes and Mathematical Statistics*. 1995, 267 pp. ISBN 0-7923-3764-6

L. Zhengyan and L. Chuanrong: *Limit Theory for Mixing Dependent Random Variables*. 1996, 352 pp. ISBN 0-7923-4219-4

A.V. Ivanov: *Asymptotic Theory of Nonlinear Regression*. 1997, 333 pp. ISBN 0-7923-4335-2

E.M.J. Bertin†, I. Cuculescu and R. Theodorescu: *Unimodality of Probability Measures*. 1997, 264 pp. ISBN 0-7923-4318-2

A. Swishchuk: *Random Evolutions and Their Applications*. 1997, 215 pp. ISBN 0-7923-4533-9

V. Kalashnikov: *Geometric Sums: Bounds for Rare Events with Applications*. Risk Analysis, Reliability, Queueing. 1997, 283 pp. ISBN 0-7923-4616-5

MIX
Papier aus verantwortungsvollen Quellen
Paper from responsible sources
FSC® C105338

If you have any concerns about our products,
you can contact us on
ProductSafety@springernature.com

In case Publisher is established outside the EU,
the EU authorized representative is:
Springer Nature Customer Service Center GmbH
Europaplatz 3, 69115 Heidelberg, Germany

Printed by Libri Plureos GmbH
in Hamburg, Germany